THE PONY FISH'S GLOW

THE PONY
FISH'S GLOW

And Other Clues to Plan and Purpose in Nature

GEORGE C. WILLIAMS

BASIC
BOOKS

A Member of the Perseus Books Group

The Science Masters Series is a global publishing
venture consisting of original science books written
by leading scientists and published by a worldwide
team of twenty-six publishers assembled by John
Brockman. The series was conceived by Anthony
Cheetham of Orion Publishers and John Brockman of
Brockman Inc., a New York literary agency, and
developed in coordination with Basic Books.

• • • • • • • • • • • • • •

The Science Masters name and marks are owned by
and licensed to the publisher by Brockman Inc.

• • • • • • • • • • • • •

• • • • • • • • • • • • •

Library of Congress Cataloging-in-Publication Data

Williams, George C. (George Christopher), 1926–
    The pony fish's glow : and other clues to plan and
purpose in nature / George C. Williams. — 1st ed.
        p.   cm.   — (Science masters)
    Includes index.
    ISBN 0-465-07281-X (cloth)
    ISBN 0-465-07283-6 (paper)
    1. Adaptation (Biology) 2. Evolution. 3. Human
evolution. I. Title. II. Series: Science masters
(New York, N.Y.)
QH546.W556   1997
576.8—dc21   96-51611

• • • • • • • • • • • • •

98 99 00 01 ❖/RRD 10 9 8 7 6 5 4 3 2 1

CONTENTS

PREFACE AND ACKNOWLEDGMENTS

The subtitle of this book might just as well have been *The Adaptationist Program,* which would have informed biologists of its subject matter. The one I chose was inspired by George Gaylord Simpson, a distinguished student of fossil mammals and one of the giants of twentieth-century biology. In January 1947 he gave an address at Princeton University, "On the Problem of Plan and Purpose in Nature," an expanded version of which was published the following June in *The Scientific Monthly.* I have no idea how either was received at the time, but I do know that they have been neglected ever since. I did not learn about them until 1965, despite many years of persistent interest in the topic and immense admiration for the author. When I finally came across it, I found the article to be, as expected, a superbly reasoned and gracefully presented review of earlier work, and a thorough demolition of much of the muddled thinking that was still prevalent on the topic of adaptation.

But I was not wholly satisfied with Simpson's treatment of biological adaptations, for which he found *plan and purpose* appropriately descriptive. He discussed the mechanisms by which organisms solve the problems of life, which do indeed seem well planned and for obvious purposes, but there is more to the *problem of plan and purpose* than that. The adaptations of organisms also show gross deficiencies in their basic plans. I hope that this book gives a balanced perspective, showing both the power and the limitations of the evolutionary process.

......

I am grateful to Cambridge University Press for permission to reprint the figure on page 26 and to Oxford University Press for the one on page 140. William Yee of the State University of New York at Stony Brook drew the pigeons pictured on page 23. The final form of that figure and most of the others was produced by Karen Henrickson of Stony Brook.

Many people gave generously of their time and effort in helping with the writing. My wife, Doris Calhoun Williams, deserves special thanks for valuable advice on the entire text, and whatever accuracy can be found in the endnotes can be credited to her. Helena Cronin also read the whole work and offered many helpful criticisms and suggestions. Margie Profet read and gave detailed advice on the first five chapters and offered much discussion helpful for the final four. Michael Ruse helped immensely with the first and last chapters. None of these good friends will be surprised that I did not always follow their advice.

··

PLAN AND PURPOSE IN NATURE

Most of us have an intuitive and adequate understanding of *plan* and *purpose* in man-made objects. The question "What is the purpose of a pencil?" need never be posed. A pencil's size, shape, material composition, and a long list of other features conform closely to an ideal design for a writing instrument. Calling it a writing instrument summarizes these descriptive details without committing us to any belief about the origin or history of pencils. They could have been invented by Rube Goldberg or by a Neanderthal. We tend to have the same attitudes about the purpose of human body parts. The idea that our ears are for hearing, for example, can be shared by people who disagree on how or when the human ear acquired its admirable design as an auditory instrument.

I am sure that pencils had a complex evolution from a crude beginning, and that changes over the centuries arose from two sources: human imagination and human experience. Imaginative inventors proposed that some modifications might be improvements, and they tried them out. The changes that really turned out to be improvements were *selected* for manufacture and use; those that did not were discarded and forgotten. In this way pencils evolved by a combination of prior plan and subsequent selection based on trial and error.

A modern biologist recognizes no element of prior plan in the origin and evolution of the human ear. Ears and other features of living organisms, like the photophore of the pony fish discussed in chapter 1, are perfected entirely by the trial-and-error process of *natural selection* proposed by Charles Darwin in 1859. Ears are maintained and improved because individuals with better ears are more likely to survive and pass their genes on to future generations. This conclusion is supported by evidence that organisms can have sophisticated adaptations and at the same time show design features that would not be there if intelligent planning had played a role (see especially chapters 1, 3, 8, and 9). The idea that human adaptations arose entirely from blind trial and error has serious implications for any honest view of human nature and the present human condition (chapters 6 to 9).

In this book I assume the validity of what has been called the *adaptationist program* in recent technical literature. For each attribute of an organism, the program's practitioners raise the question: How does this relate to the organism's efforts to survive and pass on its genes? Such a question about human teeth, for example, has an obvious answer: they play a positive role in human nutrition, which is clearly needed for survival and reproduction. But the answer is less obvious with a more specific question: Why are there four incisors in each jaw? Serviceable dentitions with three or five incisors can easily be imagined. The answer may be purely historical: primates early in their history gradually changed from some larger number to four, and all primates today are stuck with it because there is no easy way to evolve from four to three or five.

Another question might be, What is the purpose of the sounds we make when gnawing bones or biting celery? The answer: none. Noise is an unavoidable cost of the use of such mechanical adaptations as human teeth. The scientific value of any of these answers is that they have implications that can be checked. They sometimes enable us to predict and make important discoveries. I hope that the early chapters of this book provide abundant justification for this claim.

The idea that theories about prior history can be predictive is often casually dismissed. This is because people think of prediction in terms of future history, when the important use of theory is to predict the outcome of investigations. This is nicely illustrated by a great triumph of nineteenth-century science, the discovery of Neptune. Two scientists independently in France and England, using observed anomalies in the orbit of the planet Uranus, predicted that a detailed examination of a certain part of the sky would reveal a previously unknown planet. When they carried out the investigation and found the planet, this was not the prediction of a future planet, or of any future event, but merely a prediction of what would be found when and if certain actions were carried out.

The same is true of theories of human history, as illustrated by the celebrated (and controversial) discovery of Troy. A story conceived from Homeric epics and classical scholarship inspired a theory by the amateur archaeologist Heinrich Schliemann. He proposed that if he made a certain kind of investigation at a site near the western end of the Dardanelles, he would find remains of the legendary city of Troy. He carried out this investigation in the 1870s and verified his prediction. So Schliemann's theory, a narrative history with a specific geography, led to a discovery of great interest. This is routinely true of the theoretical stories told by evolutionary biologists.

·······

The first five chapters of this book summarize my view of what the study of biological adaptation is like today. An adaptation is, by definition, something functionally effective that arises from the long-continued action of natural selection. A good example is the special light of a pony fish, an admirably sophisticated aid for solving an immensely important problem. But look more closely at that fish. It has only two eyes; would it not make better sense, from a functional perspective, for it to have more than that? Its mouth and pharynx do a strange kind of double duty: they serve for both feeding and respiration. Why should the respiratory and digestive systems be associated? There is, in fact, a good reason for them *not* to be associated: the double-duty pharynx makes it possible for pony fish, and vertebrates in general, to choke on food.

And what proportion of the pony fish population would you expect to be male? I expect that most readers, like most biologists, expect the answer to be close to half. Yet surely the population's reproduction would be more efficient if only a minor fraction were male. These functionally onerous products of pony fish evolution deserve equal time with the functionally adaptive, and I hope that this book achieves something of the needed balance. The unfortunate aspects of evolution are emphasized in my final four chapters, which discuss their implications for contemporary human life—social, medical, and philosophical.

A book this size can give only the sketchiest outline of both current understandings and their implications. I hope that the endnotes will lead the more ambitious readers to other works that can elaborate on what is only outlined here.

··

CHAPTER I

ADAPTATIONIST STORYTELLING

Consider the following pair of propositions: the Sun exists to illuminate the surface of the Earth; we have eyes to enable us to make use of the sunlight. Both statements imply a cause-effect relationship. The Sun is a cause of periodic brightness on the Earth's surface, and eyes cause vision in animals that have them. Both also imply something more: that the Sun is there to fulfill a need for terrestrial illumination, and that we have eyes because we need to see. The point of this chapter is that the first of these further implications is false, or at least has no evidence in its support, and that the second is true, in a special and immensely important sense.

An examination of the Earth-Sun system utterly fails to support the idea that the sun exists to serve the planet. The sun is about 150 million kilometers away, a distance of nearly 12,000 Earth diameters. The Earth is nearly spherical and about 12,600 kilometers in diameter. Why would something that exists to serve the Earth be so far away from it? And why would it be enormously larger than what it is there to serve?

The Sun's diameter is about 100 times that of the Earth, its volume roughly a million times greater. The whole gigantic surface of the Sun is brilliantly radiating in all directions.

The Earth's small size and great distance enable it to intercept less than a billionth of the Sun's light. The rest radiates out in other directions, with other bodies in the solar system also intercepting minuscule proportions of it. The efficiency of the Sun's use of energy in illuminating the Earth is microscopic. Indeed, a detailed examination of the Sun fails to disclose any features that relate specifically to the Earth.

What should we expect of a system really designed to illuminate the Earth? Given the constraint of having to use a single radiating sphere as the light source, we might want to economize on energy and materials by making the light source much smaller than the Earth but in a close circular orbit around it. This was the standard conception in antiquity, for example, the Greeks' solar chariot crossing the sky from east to west. Even though this system's efficiency might be a million times what we now have, it would still be low from an engineering perspective. Most of the light would miss the Earth and go off in other directions. With a precisely shaped and brightly polished reflector mounted behind the Sun, we could make do with a much weaker source and achieve an efficiency to satisfy rather stringent engineering demands.

But why the constraint of a spherical light source radiating in all directions? Why not have the Earth surrounded by a grid of fluorescent tubes, or something analogous on a colossal scale, with the tubing backed up by precisely shaped reflectors? Or you could have the light produced by terrestrial objects, like the two brilliant Trees of Valinor that for long ages furnished all needed light in J. R. R. Tolkien's Middle Earth from leaves that shone from their undersides. Any such engineered light source would clearly show, by its obvious engineering, that lighting the Earth was its *raîson d'être*. The real Earth-Sun system shows no such evidence of purposive engineering.

What about the eye? This in fact is the classic example of plan and purpose in nature. It forms a centerpiece for

William Paley's version of the theological "argument from design" in his renowned early-nineteenth-century book *Natural Theology:*

> Observe a new-born child first lifting up its eyelids. What does the opening of the curtain discover? The anterior part of two pellucid globes, which, when they come to be examined, are found to be constructed upon strict optical principles; the self-same principles upon which we ourselves construct optical instruments. We find them perfect for forming an image by refraction composed of parts executing different offices; one part having fulfilled its office upon a pencil of light, delivering it over to the action of another part; that to a third, and so onward; the progressive action depending for its success upon the nicest and minutest adjustment of the parts concerned; yet these parts so in fact adjusted, so as to produce, not by a simple action or effect, but by a combination of actions and effects, the result which is ultimately wanted. And forasmuch as this organ would have to operate under different circumstances, with strong degrees of light, and with weak degrees, upon near objects, and upon remote ones; and these differences demanded, according to the laws by which the transmission of light is regulated, a corresponding diversity of structure; that the aperture, for example, through which the light passes, should be larger or less; the lenses rounder or flatter, or that their distance from the tablet, upon which the picture is delineated, should be shortened or lengthened: This, I say, being the case, and the difficulty to which the eye was to be adapted, we find its several parts capable of being occasionally changed, and a most artificial apparatus provided to produce that change. This is far beyond the common regulator of a watch, which requires the touch of a foreign hand to regulate it; but it is not altogether unlike Harrison's contrivance for making a watch

regulate itself, by inserting within it a machinery, which, by the artful use of the different expansion of metals, preserves the equality of the motion under all the various temperatures of heat and cold in which the instrument may happen to be placed. The ingenuity of this last contrivance has been justly praised. Shall, therefore, a structure which differs from it, chiefly by surpassing it, be accounted no contrivance at all? or, if it be a contrivance, that it is without a contriver?

Modern physiologists would be entirely in agreement with Paley's description of the structure and regulatory capabilities of the human eye, and would be able to add further facts in support of Paley's reasoning. He well understood the eye as an optical instrument, up to the point of the formation of a precise two-dimensional image on the retina. Today we have, in addition, a detailed knowledge of retinal photochemistry and the way photic reactions in the rods and cones are efficiently transduced into nerve impulses. We partly understand the routing of these impulses through layer upon layer of information-processing machinery in the retina itself and then in the brain, so as to achieve the maximum possible amount of useful information from the light that reaches our eyes.

But what of Paley's final question? Must a contrivance have a contriver? His asking it makes clear that he assumed only one possible answer. This is the point of his famous parable about finding a watch on the ground. Examining the watch showed it to be an elaborate contrivance for telling time. There must therefore have been a contriver (watchmaker) who understood the need to measure the passage of time and knew how to contrive a way to meet that need. For Paley, a Christian clergyman, there must be an eyemaker, the omniscient creator recognized by Judeo-Christian theology.

A CLOSER LOOK AT PALEY'S ARGUMENT
·······

Unfortunately for this aspect of Paley's reasoning, not all features of the human eye make functional sense. Some are arbitrary. To begin at the grossest level, is there a good functional reason for having two eyes? Why not one or three or some other number? Yes, there is a reason: two is better than one because they permit stereoscopic vision and the gathering of three-dimensional information about the environment. But three would be better still. We could have our stereoscopic view of what lies ahead plus another eye to warn us of what might be sneaking up behind. (I have more suggestions for improving human vision in chapter 7.) When we examine each eye from behind, we find that there are six tiny muscles that move it so that it can point in different directions. Why six? Properly spaced and coordinated, three would suffice, just as three is an adequate number of legs for a photographer's tripod. The paucity of eyes and excess of their muscles seem to have no functional explanation.

And some eye features are not merely arbitrary but clearly dysfunctional. The nerve fibers from the retinal rods and cones extend not inward toward the brain but outward toward the chamber of the eye and source of light. They have to gather into a bundle, the optic nerve, inside the eye, and exit via a hole in the retina. Even though the obstructing layer is microscopically thin, some light is lost from having to pass through the layer of nerve fibers and ganglia and especially the blood vessels that serve them. The eye is blind where the optic nerve exits through its hole. The loose application of the retina to the underlying sclera makes the eye vulnerable to the serious medical problem of detached retina. It would not be if the nerve fibers passed through the sclera and formed the optic nerve behind the eye. This functionally sensible arrangement is in fact what is found in the eye of a squid and other mollusks (as shown in the figure below), but our eyes, and those of all other vertebrates, have the functionally stupid upside-down orientation of the retina.

Paley did not really confront this problem. Little was known about mollusks' eyes at the time, and Paley merely treated the blind spots as one of the problems the eye must solve. He correctly noted that the medial position of the optic nerve exits avoids having both eyes blind to the same part of the visual field. Everything in the field is seen by at least one eye. It might also be claimed that the obstructing tissues of the retina are made as thin and transparent as possible, so as to minimize the shading of the light-sensitive layer. Unfortunately there is no way to make red blood cells transparent, and the blood vessels cast demonstrable shadows.

What might Paley's reaction have been to the claim, which I will elaborate in the next chapter, that mundane processes taking place throughout living nature can produce contrivances without contrivers, and that these processes produce not only functionally elegant features but also, as a kind of cumulative historical burden, the arbitrary and dysfunctional features of organisms?

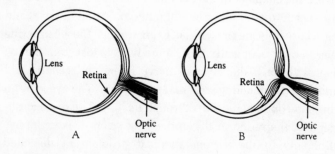

A. The human eye as it ought to be, with a squidlike retinal orientation. B. The human eye as it really is, with nerves and vessels traversing the inside of the retina.

OTHER EXAMPLES OF FUNCTIONAL DESIGN
·······

The hand is another good example of the concept of functional design. The contraction of muscles in your forearm can lead to functionally precise actions by your hand because of an elaborate system of tendons slipping through well-lubricated channels in the wrist and hand and attaching to the functionally appropriate locations on each finger bone. The opposable thumb is one of the key adaptations that led to the human mastery of tool making and manipulation. And note the fingernails: their location, their shapes, and their mechanical properties. The Roman philosopher Galen wrote a superb treatise on the human hand and other body parts, and argued forcefully for its precision and effectiveness as an instrument for manipulation. He maintained that it was so perfect for this role that it was impossible to think of any change that would improve it. Perhaps he should thus be recognized as the originator of the *optimization* concept prominent in contemporary biology.

To Galen and Paley and most scholars prior to Darwin, the functional design of living organisms led inescapably to the conclusion that there must be a wise creator who designed the organisms and their many well-engineered parts. This line of reasoning ignores an important point in analogies like that between eye and camera. Cameras surely have designers—people wise in optics and mechanics and photochemistry and perhaps economics. Yet George Eastman did more than merely design cameras on the basis of his admirable understanding. He also tried them out, tested variations in design to determine what really worked well and what less well, even if he did not always understand why. Progress in photography depended not only on engineers' understandings, but also on the data they gathered in laboratory and field tests.

What would Paley's reaction have been to the suggestion

that the creator's wisdom is as finite as ours, and that the engineering perfection of such instruments as the eye and the hand depends, like the camera, on much trial-and-error tinkering that supplemented the creator's limited understanding? And what about the suggestion that the creator had no understanding at all, but accomplished sophisticated engineering entirely on the basis of trial and error? This is essentially what is implied by modern Darwinism, as discussed in the next chapter.

The role of trial and error in the engineering of the devices that we make and use is perhaps more clearly illustrated by something simpler than a camera. Consider, for example, a fishhook. Archaeologists have found fishhooks that date from the Old Stone Age, perhaps as much as fifty thousand years ago. The early specimens were carved from bone or the curved edges of shells into a roughly hook-like form. Perhaps people carved wooden ones even earlier, but these would not have been preserved. Some time later a thoughtful individual may have imagined that a barb behind the hook's point might make it less likely to slip out of a fish's mouth. This and other advances must have been greatly aided by improvements in the working of metals. The historical details are likely to remain obscure, but barbed metal fishhooks were in use about twenty thousand years ago.

Things obviously did not stop there. Fishhooks today come in a great variety of shapes and sizes. Some have more than one barb or more than one point, often three turned 120° from one another. They vary in size, from those designed to take large sharks to those used in fly fishing for miniature trout. They vary in materials used and in the length of the shank, the curvature of the hook, and the placement of barbs. These variations did not all arise in final form from some contriver's inspiration. Fishermen using the hooks noted, from long experience, that some variants worked better than others for different fish and different circumstances. The evolution of fishhooks has undoubtedly been much influenced by a selection process. Those variants found to catch more fish were more likely to be made (or ordered) than those that caught fewer—a process that takes place with or without understanding. If a hook with a 20-millimeter shank was more reliable than one with a 25-millimeter shank, it would be favorably selected. There was no need to understand why 20 was better than 25.

I know of no way to assess the relative importance of intelligent design versus the trial-and-error process of artificial selection in the evolution of fishhooks, but surely both played a role. This is no doubt true of all the implements we use: cameras, cars, computers, and even the watch that Paley reasoned must have had an intelligent designer. How far is it possible to go with trial and error alone? All the way to the human eye and hand and immune system and all the other well-engineered machinery by which we, and all other organisms, solve the problems of life. At least that is the orthodox view in modern biology.

Darwin was challenged repeatedly on this matter. Critics would point to the precision and design features of the eye and claim that an organ of this perfection could not possibly have been produced by an accumulation of small changes, each of which made the eye work slightly better. A grossly imperfect eye, which could be improved by this process, would supposedly never evolve in the first place. Slight improvements in one part, such as the retina, would be useless without an exactly matching improvement in another, such as an increased precision of the lens. This is an utterly fallacious kind of reasoning. An improved retina may be useless without an improved lens, but both retinas and lenses are subject to individual variation. Some of the better retinas would be found in individuals who also had better lenses, so that the improvements, on average, could be favored.

The criticisms were also factually erroneous, and their proponents were ignorant of biology. As Darwin pointed out, familiarity with the animal kingdom shows the existence today of just about every stage in a plausible sequence from primitive light-sensitive cells on the surfaces of tiny worm-like animals, through the rudimentary camera eyes of scallops, to the advanced optical instrumentation of squids and vertebrates. Every stage in this sequence is subject to variation, and every stage is clearly useful to its possessor.

HOW THE PONY FISH GOT ITS LIGHT
......

Productive use of the idea of functional design, in modern biological research, often takes this form: an organism is observed to have a certain feature, and the observer wonders what good it might be. For instance, dissection and examination of a pony fish shows it to have what looks like a light-producing organ, or photophore, and even a reflector behind it to make it shine in a specific direction. So we accept the conclusion that the organ is good at producing light, but the obvious question then becomes, What good is light? The pony fish photophore is deep inside the body. Can it really be adaptive for a fish to illuminate its own innards?

The organ is situated above the air bladder, and the light shines downward through the viscera. The pony fish is small and its tissues are rather transparent. Some of the light gets through and produces a faint glow along the ventral surface. But what is the use of a dimly lit belly? Perhaps it makes the pony fish more difficult to see in the special circumstances in which it lives. It inhabits the open ocean, where it may move toward the surface as darkness approaches, but spends the daylight hours far below at depths where the light is exceedingly dim by our standards, detectable only as a murky glow from above.

Now imagine yourself lying supine in a lounge chair on

your patio looking up at the sky in a snowstorm. The sky is the leaden gray expected of this kind of wintry weather. The snowflakes are *dark* silhouettes against those stormy clouds. Snow is the ultimate standard of whiteness, as everyone knows, but pure, untrammeled snow against a gray sky looks darker than the gray. This is because that dingy sky is the source of light, and you are seeing only the shadow side of the snowflakes. Nothing between a viewer and the source of light can appear lighter than the source. This is as true of snowflakes as it is of fish. With very few exceptions, fish are darker above than below, and most have extremely white bellies. This is generally recognized as adaptive counter-shading. That kind of color distribution makes the fish harder to see from any direction. Yet no matter how white a fish's belly may be, it will appear dark if seen from below against a brighter surface.

This is true in the brightly lit shallow waters we normally encounter when fishing or swimming. It is true also in the pony fish habitat, where there is no place to hide; in all directions there is nothing but open water. The pony fish is never in bright light. It stays in water dark enough to make it unlikely to be seen by its enemies, but if it is seen it will be from below. If there is any light at all coming from above, a predator below would see a pony fish as a silhouette moving across that overhead light—unless, of course, the potential prey can extinguish the silhouette by making its belly glow in a way that matches the light coming from above.

This is just what was found in the experiments that solved the mystery of the pony fish. In an aquarium in a room with no light other than a faint glowing from above, the fish turned its light on so as to obliterate its silhouette to any eyes at a lower level. Its light closely matches the inten-sity and spectrum of the downwelling light from an open-ocean surface far overhead. J. W. Hastings did this work in the 1960s (reported in an article called "Light to Hide By"), and it is now generally assumed that this is a common adaptation in many species of open-ocean fishes, even those that have discrete external photophores.

The philosopher K. R. Popper once said: "we can describe life, if we like, as problem solving and living organisms as the only problem solving complexes in the universe." Some might object that computers are not living organisms but can solve problems. Popper, I suspect, would look at it differently, and claim that computers are merely among the means that human organisms use to solve problems.

The distinguished biologist Ernst Mayr once observed that, over the centuries, every advance in physiology began with the question, "What is the function of a given structure or organ?" An understanding of cardiac physiology was achieved by proposing that the heart is for pumping blood. Understanding a plant's reproductive physiology was achieved by proposing that its flowers function to distribute its own pollen and gather pollen from other individuals. The question "What is its function?" led Hastings to an understanding of the anatomical position, spectral emission, timing of action, and other aspects of the pony fish photophore, and of its role in protecting the fish from predation.

So adaptationist storytelling is an effective device for making important scientific discoveries, but isn't there a more aesthetic sort of triumph here? Is it not pleasing to know that nature is full of examples of such beautifully effective machinery as the human eye and hand and the pony fish's light to hide by? And is it not satisfying to know that human ingenuity can explore and understand these marvels?

Not all biologists agree with this view of the value of the "What is its function?" question. Stephen Jay Gould, for example, has drawn parallels between adaptationist theorizing and the fanciful children's "Just-So" stories of Rudyard Kipling, such as "How the Camel Got His Hump." My account of Hastings's work here might be called "How the Pony Fish Got His Light." But my story and Kipling's differ in important ways. Right from the start, mine was consistent with the known facts about the pony fish and its habitat, and I was careful to make it consistent with what might be called the Darwinian constraints: refer only to well-established material processes in formulating the story. This means avoiding any supernatural factors and always including a way in which natural selection (discussed in detail in chapter 2) could maintain the proposed adaptation.

There may be an analogy between adaptationist explanations and Kipling's "Just-So" stories, but there is a closer one with detective stories. Sherlock Holmes's only recorded trip through the American continent (in Mark Twain's "A Double-Barreled Detective Story") put him in a mining camp in Nevada in October 1900 at the time of a murder. As expected, he solved it on the basis of subtle reasoning from what seemed to others in the camp to be irrelevant details and unrelated circumstances. His solution consisted of a story that he made up to rationalize these various clues. The story proposed that one of the miners had killed the victim by blowing him up with powder used for blasting in mines. The motive was to silence the one eyewitness to a crime. The great detective's story did more than just rationalize the known facts. It predicted one not yet known, and checking on this additional fact provided all the proof that anyone needed that Sherlock Holmes's story indeed represented the truth. He had noted some blood at the scene of the crime and inferred that it had come from the murderer as a result of the explosion being more forceful than planned. The murderer would be known by a wound with just the sort of location and severity that would explain the observed blood. Everyone looked around the tavern in which the great detective was telling his story, and there stood a man with "the blood-mark on his brow."

Everyone was satisfied, and awed by the brilliance of the great visitor, except a certain Archy Stillman, who had the audacity and poor taste to challenge the authoritative pronouncement. Stillman proposed a contrary story: the murder had been committed by a vengeful assistant to the victim. This story fit all the facts that Sherlock Holmes's had, and a few others, and it also predicted that a search near the scene of the murder would disclose some of the discarded equipment, belonging to the assistant, that had been used to produce the explosion. At this point the assistant hysterically confessed, verified that he had used the equipment exactly as the story described, and resigned himself to speedy capital punishment.

Holmes's theory was superseded by another theory, in a way similar to what happens in science. A newer theory that more neatly explains the known facts and accurately predicts new ones is accepted and an older one thrown out. As pointed out by James B. Conant in 1947, the reason for scientific theories being abandoned is always that a better theory has come along. Theories are never abandoned merely because there seem to be some contrary facts; proponents of a theory can always find excuses for such discrepancies.

Similarly, there could be more than one story to explain how the camel got his hump or how the pony fish got his photophore. The one that best explains the facts is the one that will be accepted. In science as in sleuthing, acceptability also depends on adherence to accepted rules for storytelling. It is not proper to invoke processes not known to be generally applicable to problems like the one to be solved. Divine intervention, for instance, may not be postulated, either by biologists or detectives. The explosion that killed the unfortunate miner could not have been from a thunderbolt hurled by Jupiter, nor can the camel's hump be attributed to a god directing the course of evolution. Hastings's pony fish story thoroughly conformed to the rules of the game: it invoked only known processes operating in the fish and its habitat, and, like the alternative solutions to the murder mystery, it was consistent with the known facts.

Also, like both Holmes's and his challenger's proposals, it

had implications not known to be true. Fortunately, some of these statements could be checked, and they thereby constituted predictions of what various investigations would reveal. The predictions were in fact supported by Hastings's inquiries, and can be cited as evidence for the validity of his story, his solution to the mystery of what the light in the pony fish was for. He verified predictions about when the light would be turned on, what kind of light it would emit, and other features of pony fish biology that would not have been discovered without his adaptationist story about that fish.

In another important respect, analogies between adaptationist explanations and Kipling's tales or the solution to murder mysteries can be misleading. Hastings's story is not really about how the pony fish got its photophore but rather *why* the pony fish *keeps* its photophore. It says nothing about what the photophore was like ten thousand or ten million years ago. Adaptationist stories are not about evolution so much as about its absence. The pony fish may now be evolving rather rapidly on the time scale of historical geology, on which a million years is a short time. I also suspect that it is capable of evolving perhaps a thousand times as fast as it actually is. Domesticated animals and plants routinely evolve a thousand times as fast as they did in ancestral habitats. But evolutionary rates and directions are irrelevant to the adaptationist explanation of why the pony fish keeps its photophore, which deals solely with the current utility of this remarkable organ.

Gould's criticism notwithstanding, adaptationist storytelling continues to be a powerful method for the discovery of important facts about living organisms.

....

CHAPTER 2

FUNCTIONAL DESIGN
AND NATURAL SELECTION

In 1859, when Darwin published *The Origin of Species*, the idea of evolution was very much in the air. Scientists generally recognized that fossils are the petrified remains of creatures long dead and often extinct, some strikingly different from anything living today. They also recognized that the plants and animals they knew were totally absent from fossil assemblages in many rock layers. Life forms had apparently changed through the ages, and explanations for why this should be were various. The *catastrophist* school of thought assumed that the strange organisms of earlier ages had all been destroyed in great calamities, with other plants and animals created to replace them after each calamity. The renowned French biologist Lamarck thought that current animals and plants had evolved slowly from earlier forms. He envisioned evolution occurring partly as a kind of predetermined developmental process and partly from the compulsive strivings of the organisms themselves.

Physical scientists studying rock formations were also devising evolutionary theories for what they observed in the Earth's crust. They came to believe that some rocks form from the slow consolidation of sediments that gradually accumulate and that others form in other ways, for instance, from hot volcanic masses forcing their way through overlying rocks or flowing out onto the surface. They understood that the relative ages of adjacent rocks might be inferred. Younger sediments usually lie on top of older ones, and a volcanic intrusion must be younger than the rocks through which it flowed. Absolute ages were more in doubt, but calculating how long it would take known processes to produce observed results suggested that the Earth must be far older than would be allowed by biblical reckonings.

Perhaps the most noteworthy of these pioneers was the Scotsman James Hutton, regarded by some as the founder of historical geology. His writings implied a possibly infinite age for the Earth, which he envisioned to be in a slow but unending state of repetitive upheaval. In 1785, a quarter-century before Darwin's birth, he maintained that an objective and comprehensive examination of crustal rocks revealed "no vestige of a beginning, no prospect of an end." So the idea of an immense amount of time available for evolutionary change was intellectually respectable in Darwin's time, and it implied that even very slow evolutionary processes might bring about great changes.

DARWIN'S THEORY OF EVOLUTION
•••••••

The slow process that Darwin proposed as most important in evolution was *natural selection*, a process deduced from two abundantly supported generalizations. The first is that there is a struggle for existence throughout the living world. In every species of plant or animal, more individuals are produced in every generation than can possibly survive and reproduce. Some will succeed, others fail. The second generalization is that there is such a thing as heredity. Offspring tend to resemble their parents more than they do other adults of the parental generation. Darwin reasoned that such variation could affect characters important in the struggle for existence, and he found many examples of this sort of variation. It follows that each generation will have a biased representation of the variations found in the preceding one. Whatever helped their parents in their struggle for existence will be more abundantly represented in the surviving offspring than traits that handicapped individuals in the parental generation.

He supported this theory by a massive array of evidence from natural history and by analogies with artificial selection. Breeders, in choosing individual plants and animals for breeding stock, usually select those with the features they like best. Those less well endowed are sold or eaten or otherwise eliminated. By this sensible practice they can induce gradual change over many generations, so that domesticated forms often look and act quite different from their wild ancestors. Today, after thousands of years of selective breeding under domestication, we have breeds of dogs and horses and roses and strawberries that are quite different among themselves and from the wild species from which they were derived.

Darwin himself bred pigeons and used the origin of pigeon breeds as a model for the origin of diverse stocks from a single ancestral species. The illustration shows some of the diversity of pigeon types, all produced by rapid evolution under domestication.

Darwin argued that if farmers or hobbyists, by frequently culling the least valuable of their pigeons or pigs or potatoes, can produce varieties that are economically or aesthetically superior to the ancestors, nature can surely do something similar. Competition and adverse conditions of life impose an automatic culling process in every generation. The result should be that the wild animals and plants develop ever greater ability to survive this culling process. The philosopher Herbert Spencer later called this principle *the survival of the fittest*, a handy if somewhat misleading phrase.

A sampling of the diversity of domestic pigeons, all derived from the wild Eurasian rock dove (center), through recent centuries of selective culling by breeders. Compare the diversity here with that of Darwin's finches (p. 26), for which hundreds of thousands, perhaps millions of years were available.

Darwin argued that if this process operated through enormous numbers of generations and, especially, if the environmental conditions that caused the culling changed from time to time, major evolutionary modifications would be expected. Descendants of a single species of ancestor, if they inhabited different regions subject to different conditions, could diverge to such an extent that they would have to be considered different species from the ancestor and from each other. One of his examples from nature was the finches of the Galápagos Islands, a small archipelago on the equator a thousand kilometers west of South America. The islands rose from the sea as volcanoes a few million years ago, and were never a part of the mainland. They were largely inaccessible for land animals and plants, and most of the common inhabitants of South America did not exist there when the islands were first explored. The absence of things such as frogs and small mammals (other than bats) is easily explained: the great expanse of open ocean is a forbidding barrier, even for birds of the tropical American deserts and forests. Yet it should not be surprising that some limited colonization by land birds did take place as a result of accidental straying from the closest continent.

Darwin, in his visit to the islands as naturalist aboard the research vessel *Beagle,* found the descendants of such colonists, about a dozen species of finch whose closest relatives were in South America. He noted that each major island had one or more of the distinct species. He theorized that sometime after the islands had formed and become habitable, some South American finches, conceivably just one of each sex, reached one of the islands. Perhaps they had been caught in a storm, with an easterly wind blowing faster than they could fly, and lucked upon this isolated land instead of dropping exhausted into the sea. They survived and bred in their newfound home, where competitors were absent and conditions at least minimally met their requirements for food and nest sites. Soon, perhaps in just a few years, they built up a large population, then some of them occasionally reached another of the islands in the archipelago to repeat the process.

But why should there be so many species of Galápagos finch, and not just the one original colonist? Darwin answered this question by noting the differing environmental conditions on different islands. Some are large, comparable in size to Rhodes or Minorca, others smaller than the one that supports the Statue of Liberty. Some have high mountain peaks that catch considerable rain, others are low and dry. The diversity of conditions produces a diversity of vegetation and of the seeds and insects that finches feed upon. These differing environmental circumstances demanded different capabilities in the finches' struggle for existence.

Of special importance is food suitability. Some potential food sources are seeds with formidable shells. If only some of the newly established finches were able to break the shells, powerful selection for these more powerful beaks and jaw muscles would be brought to bear. In perhaps a few thousand generations, the finch populations of different islands would show differences in their feeding adaptations. In hundreds of thousands of generations, the diversity of these features could have reached the level found by Darwin, as illustrated in the figure.

This gradual alteration of a line of descent, with different lines showing different evolutionary changes, was the central theme of Darwin's 1859 book and the inspiration for its title. One of his main sources of inspiration was the differences he often found in comparing members of the same species from different areas. His trip around the world in the *Beagle* provided abundant opportunity for comparing animals and plants across their geographic ranges.

He often found that differences shown by some organism from different parts of its range were only moderately distinct, and he recognized them as *varieties* of the same species. At other times it was not at all clear whether he was dealing with different varieties or fully distinct species. Alfred Russel Wallace, who proposed the theory of natural selection simultaneously with Darwin, entitled his publication "On the Tendency of Varieties to Depart Indefinitely from the Original Type." Today natural selection is part of the standard conceptual equipment of biology, routinely

Illustration from David Lack's 1947 book on Darwin's finches. Compare with pigeon diversity shown on page 23.

invoked to explain differences among closely related organisms living under slightly different conditions.

SEXUAL SELECTION
·······

Relentless competition was an essential premise of the theory of natural selection as proposed by Darwin and Wallace. Today biologists recognize two kinds of competitive behavior among animals: *scramble* and *contest* competition. A flock of sparrows picking up grass seeds as fast as a gardener can broadcast them would exemplify scramble competition. Another would be two or more dogs chasing a squirrel, with one catching it and eating the whole carcass. If two dogs seize the same squirrel and a tug of war ensues, we have a contest, the two dogs pitted one on one against each other. It would be even more obviously a contest if they fought over the dead squirrel, directing their attention exclusively to each other, even though the squirrel is the real focus of the dispute. It would also be a contest if the dogs merely threaten each other, without actually fighting. If one dog's threat causes another to back away, it has won the contest.

Tactically similar contests may be waged over food items, roosting sites, mates, and many other resources. In fact, there need not be a currently identifiable resource for a contest to occur. It may be waged merely to establish the winner-loser relationship itself. Later on, when something like a food item is found, the loser will concede it to the winner without further dispute. This phenomenon of stratified social structure is often seen by observers of animal behavior both in the wild and in captivity. Contests are often provoked by the appearance of an item coveted by more than one individual, but the prize being sought may often be an elevated social status, which can later be used to gain access to needed resources.

The general prevalence of competition for social status has

been recognized only recently. It was obscured for Darwin, and for generations of his followers, by the fact that the most conspicuous contests are often among males for opportunities to mate with females. This fact led Darwin to propose a special evolutionary factor, *sexual selection*, that operates in addition to natural selection. Sexual selection normally depends on contests between males, with the winner gaining and the loser forgoing sexual access to one or more females. This competition for social status often takes place without any females being present. Many male migratory birds, for example, precede females on the spring migration, and establish their social hierarchy before the females arrive.

In some animal species, one male contends directly with another, the fight between stags in the rutting season being a clear example. In other species, males contend indirectly by displaying to females a train of enormous feathers, a peacock being the classic example. The contest is still between males, the winners being those best at impressing the females. In still other species, the contest is for a territory, and waged with threats or actual combat at tentative territory boundaries. Persistent winners get larger territories in choice breeding habitats; losers settle for smaller territories or inferior habitats. The females may choose males indirectly by seeking the better places to lay their eggs. In other territorial species, a male must actively court a female and try to entice her to his own nesting site. The male threespine stickleback, an object of many classic studies of reproductive behavior and sexual selection, threatens and fights other males to secure a territory in a good nesting habitat. Then he builds a nest and must entice females to it while keeping other males away.

Darwin was led to propose sexual selection by the many conspicuous features of animals that could not be attributed to natural selection as he envisioned it. They were features more likely to hinder than to aid an organism in its struggle for survival. Again, the conspicuous nuptial train of a peacock, which is so burdensome as to make normal flight difficult, is a good example. As a general rule, these conspicuous and burdensome features characterize adult males rather than the females or juveniles of a species, and they often appear only in the breeding season.

Darwin reasoned that the great burden of ornamental plumage may make life difficult and dangerous for a peacock, but he proposed that it may be favorably selected if its display makes it easier for the male to compete for a mate. Sometimes the burdens are clearly weapons. The antlers of a male deer, grown anew each breeding season, are an obvious instance. Other sexually selected features are only for display, playing a role in courting females, threatening rival males, or both.

Sexual selection is an idea routinely invoked in biological research today. It is generally viewed as a pervasive evolutionary factor, more important perhaps than any other kind of selection in relation to contests with other members of the species. The concept has moved beyond an explanation for competitive behavior or the production of weapons or displays to encompass many physiological processes and has been extended to such phenomena as the floral displays and selective use of incoming pollen grains by plants. But during Darwin's lifetime, he was the only prominent biologist to argue the importance of sexual selection, and he used it only to explain special features of adult male animals. Wallace, who in some ways seemed a more extreme advocate of natural selection than Darwin, had little use for Darwin's theory of sexual selection. He ridiculed the idea that females could be influenced by the displays that males seemed to be directing toward them. Even as late as 1950, biologists studying animal behavior made little mention of sexual selection. In this and other ways, Darwin was far ahead of his contemporaries.

But many modern biologists would concede that Darwin erred in thinking of sexual selection as a process separate from natural selection. They regard it as a special category of selection for social status, which is a kind of natural selection. This idea implies recognizing that members of one's own species are a feature of the environment and that adaptations to this feature are expected. Social status is a kind of resource that can never be in adequate supply. The top dog in a pack has all he needs, no doubt at great cost to himself and others, but the other dogs all need more than they have and will do what they can to get it. Unlike food, an individual's social status is a resource that can never be lost to a member of another species. A ferret may compete with a fox for a rabbit, but never for social status among foxes. (Or, perhaps, almost never. The first-century horse Incititus gained a lofty rank in human society at the expense of multitudes of men of lower rank.)

Darwin achieved a high level of recognition in Victorian society as a scholar and expert on many aspects of natural history and for establishing the acceptability of evolution. Yet in retrospect we can recognize that he failed to convince many people that natural selection is the main force of adaptive change. From Darwin's death in 1882 to the 1920s, evolution, his "descent with modification," was generally accepted by biologists, but not natural selection or sexual selection as causes of the modification. Many leading scientists of this period advocated theories that today seem naive and fanciful, such as orthogenesis, the idea that evolution has some kind of momentum that keeps it going. Some continued to prefer Lamarck's ideas to Darwin's.

NATURAL SELECTION AND THE PREVENTION OF EVOLUTION
......

Paradoxically, much reference to natural selection today relates not so much to evolution as to its absence. If natural selection is the reason the pony fish keeps its photophore, then it is preventing the loss of this organ by evolutionary change. We know now, from abundant experiments on the evolutionary potential of living organisms, that they are capable of evolving far more rapidly than is normally observed today or found in the fossil record. What natural selection mainly does is to cull departures from the currently optimum development of the features shown by organisms. If some species of bird has wings that average 20 centimeters long, it is assumed that individuals with wings of 19 or 21 centimeters would have a slight disadvantage. They would be less likely to survive to maturity and would have lower survival or fertility rates thereafter. Evidence for exactly this was shown by a classic study of natural selection in the wild. In 1899 the British biologist Herman Bumpus measured the wings of a large number of sparrows that had been killed in a storm. He found that those with markedly longer or shorter wings were more abundantly represented among those killed than in the population at large.

The advantage of having intermediate character development (wing length, insulin production, coloration, and so on) is often called *normalizing selection* or *optimization*. Most of the selection taking place in nature is assumed to be of this sort, rather than any that would cause an observable shift in average values from one generation to the next. Even the weak directional selection that does take place is usually thought to be corrective. The population would evolve to be less well adapted if natural selection did not weed out occasional adverse mutations or locally maladaptive genes introduced by individuals moving in from places where conditions are different. So the process proposed by Darwin as the major cause of evolution is now thought to operate mainly to *prevent* evolution. Aristotle's descriptions of wild animals and plants, written 2,500 years ago, are still accurate for their descendants today, mainly because natural selection has been preventing their evolution. The domesticated animals and plants that Aristotle observed were often strikingly different from what farmers grow today, because artificial selection has been causing their rapid evolution.

The concept of character optimization has been with us ever since people first tried to understand the workings of their own bodies and those of other organisms. Aristotle and Galen used the idea habitually, as noted in chapter 1. In 1779 the British philosopher David Hume, in contemplating the quantitative precision of biological mechanisms, proclaimed:

> All these various machines, and even their most minute parts, are adjusted to each other with an accuracy which ravishes into admiration all men who have ever contemplated them. The curious adapting of means to ends, throughout all nature, resembles exactly, though it much exceeds, the productions of human contrivance.

Note the similarity in sentiment between this statement from Hume, an atheist, and that of the orthodox Christian Paley (see chapter 1). Both were clear-thinking and keen observers of nature.

More recently the concept of optimization has been extended to aspects of biology where its applicability is less obvious. Many recent studies of life histories and animal behavior are good examples. Biologists today speak of the optimization of egg size and number, of mate choice, of the seasonal timing of migration. Optimization is used to understand and predict such things as how long a bee will stay at one clump of flowers, how big a load of pollen or nectar it will pick up before returning to the hive, and at what times of the day it will go foraging.

It is ironic that many prominent biologists, during Darwin's time and for many decades thereafter, tried valiantly to demonstrate some force of evolutionary adaptation that could cause change more rapidly than natural selection.

They could not imagine that so weak and misguided a process as Darwin proposed could actually produce the observed complexity and diversity of life, even with liberal estimates of the amount of time available for it. Nowadays it is more fashionable to wonder what makes evolution so slow. Some organisms living today are closely similar to ancestors of more than a hundred million years ago. The orthodox reasoning is, in the words of Roger Milkman, a distinguished geneticist at the University of Iowa, that "[t]he main day-to-day effect of natural selection is the maintenance of the status quo, the stabilization of the phenotype. To a relatively small directional residue, we attribute the great panorama of evolution."

The current trend is not to doubt that natural selection could produce the great panorama but to doubt that it can account for the stability. It is proposed that selection also acts at higher levels than the loss or survival of genes in populations. The extinction of whole evolving lineages could have persistent biases that cull most newly formed groups of organisms in ways analogous to the weeding out of most mutations within evolving populations. So a natural selection among whole populations or larger groups of organisms would be, like selection within populations, concerned mainly with maintaining the status quo.

GENETICS, MOLECULAR BIOLOGY, AND MODERN DARWINISM
·······

Darwin's vague generalization that "like begets like" is an adequate premise for the basic logic of his theory of natural selection, but it does not permit many quantitative inferences. It gives no hint as to why offspring should show a resemblance to their parents. Today we have the science of genetics, with a detailed theory of heredity that allows much more rigorous thought about evolution than was possible in the nineteenth century. Histories of scientific fields are usually vague about origins, but genetics is an exception. It began decisively in the 1860s with experiments on peas grown in his monastery garden by the Augustinian monk Gregor Mendel. He published his work in 1868, but it was ignored for the rest of the century. In the early 1900s it was discovered by several biologists investigating heredity in a variety of different organisms. They belatedly recognized the profound significance of the work of that lonely scientist.

What Mendel had found, and the later workers confirmed, was that crosses between parents of strains that differ strikingly in some character will often show predictable ratios of the contrasting features in subsequent generations. A parental character, such as short stems, may disappear entirely in the first offspring generation, all of which have long stems (the *dominant* character). Crosses between these individuals will produce offspring of which about 25 percent show the (*recessive*) short stems missing in their hybrid parents. Crosses between a first-generation hybrid and the recessive strain produce a nearly equal number of long and short stems (dominants and recessives) in their offspring.

These regularities (*Mendelian ratios*) can be explained by a precisely controlled and strictly particulate theory of heredity. Today we call the inherited particles *genes*. They are particulate in that they retain their identity in passing through generations. A gene is either inherited or not, passed on or not, with never any sort of partial presence. By about 1930 it was clearly shown that the genes are in a nearly constant linear arrangement on the chromosomes, visible with special techniques in dividing cells. The chromosomes are present in pairs, with each member of each pair having the same linear arrangement of genes, one in each pair having come from the mother, the other from the father. So the paired chromosomes imply paired genes. If the gene inherited from one parent differs from that from the other, biologists refer to two different *alleles* of that gene.

When an individual forms an egg or sperm, the corresponding (homologous) chromosomes line up, exchange some corresponding parts, and then separate, each chromosome going at random to one or the other of the resulting cells. This exchange of parts and random segregation of chromosomes assures that any two alleles in an ancestor will ultimately go their separate ways in descendants. The genes pass indefinitely through the generations, but gene combinations (genotypes) are unique and fleeting, as long as reproduction is sexual. The implications are well worth bearing in mind. You got half your genes from your mother and the other half from your father, one-eighth from each great-grandparent, and so on. Each of your children got half your genes, each grandchild a quarter, and so on. You are the bearer of a legacy of genes from the past. Each allele at each locus has its own unique history, back to a possibly reomote origin by mutation from a contrasting allele. Yet your genotype never existed before you were conceived and will never be produced again.

For the first half of this century, there was great uncertainty about the genes' chemical nature. In retrospect we can say that it was obvious by the 1940s that genes can be identified with deoxyribonucleic acid (DNA). All doubt was laid to rest by the much-lauded work of James Watson and Francis Crick in 1953. They resolved the detailed chemical structure of DNA and showed how it serves as a medium of communication within a cell lineage and between the generations of multicellular organisms. Because of Watson and Crick, we now know that heredity is not only particulate but *digital*.

Other examples of digital information transfer are printed English words with their twenty-six-letter alphabet, Arabic numerals with their ten-letter alphabet, and Morse code and "computerese" with their binary alphabets. The genetic code has a four-letter alphabet of molecular structures with names abbreviated as *A*, *T*, *G*, and *C*. Any sequence of three such groups can specify a particular amino acid, one of the building blocks of proteins. For example, *C-A-G* specifies the amino acid *glycine*. If the code were changed to *C-C-G*, some protein would contain the amino acid *proline* at the position that would have been occupied by the glycine. On the other hand, changing *C-A-G* to *C-A-A* has no effect: the amino acid specified is still glycine. This is one of many examples of redundancy in the genetic code. Some different DNA sequences are functionally synonymous, just as, in English, *gray* and *grey* mean the same thing. An understanding of the DNA code is basic to an understanding of evolution.

Imagine that, in some population of some organism, a gene has, among its thousands of base pairs, the sequence *C-A-G* and has, for many generations, been reliably putting a glycine into some protein, perhaps an enzyme. The mechanism that allows this gene to be so amazingly stable will be discussed in chapter 5, but for now I will merely point out that no mechanism has absolute reliability. Rarely, the *C-A-G* may change to some other sequence, perhaps *C-C-G*, so that the resulting enzyme, in cells containing the new sequence, will have a proline in place of the glycine. This might affect the action of the enzyme to a considerable extent, or maybe only slightly, perhaps scarcely at all. If the change is an improvement, natural selection may cause the allele with the *C-C-G* to replace that with the *C-A-G* at its position on its chromosome throughout the population.

Any new mutation can be lost by chance. This is the most likely event for any new allele, even one that gives a substantial advantage. But mutations occur with finite frequencies. If *C-A-G* -> *C-C-G* has a one-in-a-million probability per germ cell (egg or sperm), and there are about a thousand individuals per generation, a mutant individual should appear about once in every thousand generations, and ten times in ten thousand—a trifle, in evolutionary terms, for many organisms. Sooner or later a favorable mutation should catch on and start replacing the ancestral allele at that locus.

The great beauty of Mendelian heredity in its evolutionary application is that it lends itself readily to quantification and precise reasoning. This is the subject matter of population genetics, a field well established by the 1930s. Population geneticists can deal with such quantities as mutation rate; frequency of recombination of genes on the same chromosome; expected rate of replacement of alleles by better-adapted mutant forms; expected levels of chance deviations from expected rates as a function of population size and other variables; differences in these rates between recessive and dominant genes; and many other influences on the evolutionary process. These quantitative variables can be related to one another algebraically and evolutionary conclusions drawn as solutions to algebraic equations.

For instance, it can be shown that natural selection can be far more powerful than we might intuitively expect, and can accomplish major changes in brief periods of evolutionary time. Imagine maintaining a herd of a thousand gray horses, with a modest level of starting variability in shades of gray and rates of new mutations affecting this character. Visit this herd once per century and remove the palest specimen. Simple calculations can show that this procedure could result in a herd of uniformly black horses well within a million years.

Recently some Swedish workers reached an even more startling conclusion. Assuming nothing more than some cells of a primitive animal with some sensitivity to light and modest rates of mutations affecting that sensitivity, the position of the cells in the body, the transparency of overlying tissues, and other relevant variables, they showed that it could take as little as 400,000 years to evolve the vertebrate eye. This is less than a thousandth of the time that has elapsed since multicellular animals first appeared. This is an especially interesting example because Darwin's critics have long cited the eye as an example of an organ that is far too complex and precise for any short-sighted process such as natural selection to produce.

Intuitions about the evolutionary process can be a great source of ideas but not of conclusions. Conclusions must be based on precise quantitative reasoning, such as realistically formulated mathematical equations or carefully designed graphic models. Such reasoning must be focused in a way that leads to testable expectations about the real world, such as what a series of measurements on a group of fossils will reveal, how an experiment on microorganisms grown in specified environments will turn out, and so on. The maintenance of proper scientific rigor is, of course, seldom easy, even for well-trained scientists.

. . . .

CHAPTER 3

DESIGN FOR WHAT?

> Now, as each of the parts of the body, like every other instrument, is for the sake of some purpose, viz. some action, it is evident that the body as a whole must exist for the sake of some complex action.
>
> —*Aristotle*

The textbook for a college biology course I took in 1947 gave the following statement of the theory of natural selection:

1. **Variations** of all grades are present among individuals . . .
2. By the **geometric ratio of increase** the numbers of every species tend to become enormously large; yet the population of each remains approximately constant because . . . many individuals are eliminated; this involves:
3. A **struggle for existence**; individuals having variations unsuited to the particular conditions in nature are eliminated, whereas those whose variations are favorable will continue to exist and reproduce.
4. A **process of natural selection** is therefore operative, which results in:
5. The **survival of the fittest**, or "the preservation of favored races."

The quotation in item 5 is from the subtitle to Darwin's *Origin of Species*. Unfortunately, Darwin was never clear about what he meant by "race." Is the appearance of a novel feather pattern in a flock of domestic pigeons the start of a new race? Or is there a new race only when the pattern is bred for and comes to characterize a large stock? Are individual differences in wild animals and plants to be considered racial differences? Or are races in nature always groups of individuals, often inhabiting different regions but recognizable as belonging to the same species? These variants of different geographic regions were often called *varieties* or *subspecies*, rather than *races*, in Darwin's time. *Subspecies* is the preferred term today.

Whatever Darwin may have meant by *races* in the subtitle of his book, the concept for the theory of natural selection as Darwin used it and as taught in 1947 was that implied in items 1, 2, and 3. It is variation among competing *individuals* that provides the raw material for natural selection. This is also clear from the textbook's subsequent detailed discussion of the theory. Other texts in use for the next two decades continued to imply that natural selection operates among competing individuals of the same neighborhood. Since the 1970s, textbooks have been more likely to be explicit on this point, and to insist that natural selection operates strongly among individuals and that selection among races or other collective entities is usually a weak influence on the course of evolution.

Moving ahead a few years, I find myself in a graduate seminar in marine ecology. The subject is the adaptations by which little fish try to avoid being eaten by big fish. One example recognized is toxic flesh. If a 10-kilogram barracuda eats a 1-kilogram poisonous perch, it may die or at least be sickened and thereby deterred from attacking that kind of prey in the future. It was agreed by the discussion leader and most of the students that poisonous flesh was a good example of a protective adaptation.

But there was one skeptic, and his name was Murray A. Newman. He would soon go on to a distinguished career as director of the splendid public aquarium in Vancouver, but that day he was a lone dissenter. "Wait a minute," he muttered. "How can being toxic protect you? It does nothing to the predator until long after you're dead." The immediate vehement reaction from me and several others: "That's stupid, Murray. The toxicity doesn't have to protect the toxic individual; it protects the species in general." There were no further objections and the discussion continued, but I am not sure that Murray was convinced. I think I was at the time, but not firmly and not for long. I would soon be increasingly nagged by the seeming inconsistency between the theory of natural selection as presented in textbooks and the "good of the species" adaptations that were routinely attributed to the process, sometimes glibly by the same texts that presented a strictly selection-among-individuals form of the theory.

WHAT IS AN ADAPTATION'S ULTIMATE PURPOSE?
....

The message of chapter 1 was that the parts of organisms are functionally well designed: the eye for vision, the hand for manipulation, and so on. But what are vision and manipulation for? They are important for a long list of vital functions, and without them life would be much more difficult. Vision and manipulation are often called on at the same time, the former helping coordinate the latter. A blind man or a man without hands would take a long time to gather enough wood for a useful fire. What is fire useful for? Cooking or perhaps just warmth. And what good are cooking and warmth?

I could go on with questions beyond questions, but perhaps not much further. Soon I would come to such values as health and useful skills and social status. What are they good for? For Aristotle's "complex action," that's what. Or at least for the modern biological interpretation of the complex action to which all adaptations contribute: reproductive success. Continued physical survival is ordinarily significant in evolution only if it increases the likelihood or extent of reproduction.

But reproduction is a tricky concept for sexual organisms like ourselves. I have children and grandchildren, but I am not really present in either generation. None of those individuals inherited any of my body parts, and none is more than 50 percent similar to me genetically. I happened only once, I will never be duplicated, and when I am gone it will be forever. My descendants' biological heritage from me is limited to a sampling of my genes. Half the genes in each of my children and a quarter of those in each grandchild came from me. As noted in chapter 2, my complete set of genes, my *genotype*, cannot be passed on. So reproduction, for a sexual organism, has a restricted meaning. It means passing on genes, dissociated fragments of the organism's genotype. A mother provides an egg and a father a sperm, and each such cell contains the single set of genes on a single set of chromosomes. The combination of egg and sperm genes provides the genotype of the new individual, which then develops according to its new and unique instructions.

It looks as if reproduction is the ultimate adaptation to which all others are subordinate, but that is too simple, because producing offspring is not the only way to get one's genes into future generations. An organism can also transmit its genes by helping with the survival and reproduction of relatives. Consider the diagram, which is meant to show genealogical connections among eight individuals. The overall ability of individual #5 to get her genes (indicated by the shading) into future generations is termed her *inclusive fitness*. The evolutionary process that maximizes the ability to treat others according to their genetic similarity to oneself is termed *kin selection*.

Statements about genetic variation and uniformity in sexually reproducing populations are often confusing because different measures are being discussed. In a statement such as "We're still 98 percent chimps in our genes," it is obviously base-pair similarity of chimpanzee and human gene pools that is being discussed. If part of some gene in a human cell reads GTTAGCC and exactly the same sequence of the chemical groups (nucleotides) is found in the same place on the same gene in an ape cell, the two are 100 percent similar for this sample. So what? A gene is made up of thousands of base pairs, not just a sequence of seven, as in the example shown. For two genes to be really the same, every one of the thousands of base pairs has to be the same.

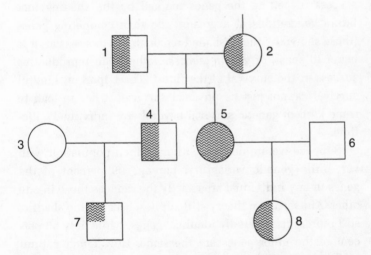

Genealogical relations in a sexual population. Squares represent males, circles females. Horizontal lines connect mates, vertical lines show their offspring. The shading indicates genes necessarily shared with individual #5. The process of kin selection would be expected to result in her treating her father (#1) as if his survival and reproduction were half as important as her own, her nephew (#7) a quarter as important (all else being equal), and so on.

If proportions of such exactly similar genes are estimated, it may be found that a quarter to a third of the genes may differ in human and ape cells. Human cells from two different individuals may well differ in several percent of their genes.

For a pedigree diagram, statements such as "a child has a 25 percent similarity to a half sibling" would have a still different meaning. Twenty-five percent of the genes are assuredly the same, because they came from the same parent. What of the other 75 percent? We do not know. Many of them might be the same. We know that they came from the same population of which the individuals are members, but that is all we know. We have to consider them a random sample of the genes in that population. The genes indicated by the shading in the diagram are special, and are technically termed genes *identical by descent*. Favorable selection of individual #5 in modern evolutionary theory relates to her success in getting the genes marked by the shading into future generations at a greater rate than competing genes (those shown by the unshaded regions). For this reason, it is better to speak of *genetic success* rather than reproductive success or the survival of the fittest. The important kind of survival cannot just be physical survival; it has to lead to some kind of genetic survival beyond any individual's lifetime.

Suppose we are dealing with an inbred population with very little genetic variability; perhaps 99 percent of the genes in any individual are exactly the same as those in any other. Kin-selection theory still applies, because the shading still represents assuredly identical genes, while only 99 percent of the other genes are the same. How much natural selection can take place in a population in which 99 percent of the genes are identical? Surely some, less than if only 90 percent were the same, but more than if 99.9 percent were. All I am doing here is applying to kin selection the same constraint that applies to any other form of natural selection. The process will not work if there is no genetic variation, and any shortage of such variability will retard the process. Without genetic variation there can be no evolution, from natural selection or any other cause.

The complete absence of genetic variation in a sexually reproducing population that numbers in the thousands or more is so improbable that it is seldom a part of the thinking of evolutionary biologists. Also, as explained in chapter 2, biologists often use natural selection to explain why an organism has the features it has rather than conceivable alternative features. They are not concerned about whether it evolved those features rapidly or slowly. So such biologists are seldom concerned about whether the organisms they study differ at 10 percent of their gene loci or at only 1 percent. Either way, they would expect the same conditions to be produced by selection in the long run, and they assume that the long run has in fact happened. They are interested not in evolutionary change by natural selection but in the evolutionary equilibria already established by this process.

THE FUNDAMENTAL IMPORTANCE OF GENETIC SUCCESS
·······

The distinction between reproductive success and genetic success is clearest in the social insects of the order *Hymenoptera* (ants, bees, and wasps). For the moment I will ignore various complications, such as the great variability of mating systems and social structures, and discuss the classic pattern in this group. A fertile female (*queen*) mates with a male and starts a new colony, with or without the help of companions from her old colony. She can lay two kinds of eggs, those fertilized by her mate and those from which she withholds fertilization. These unfertilized eggs get a single set of chromosomes and genes from the queen, none from any other source. They all grow up into males. The fertilized eggs develop into individuals with a set of chromosomes and genes from each parent, and they all become females. Most of these females grow up into sterile workers that stay with the queen and do everything for the colony except reproduce. All the workers are daughters of the queen, and none of these workers have offspring that carry their genes into future generations.

But the workers have superb adaptations for their way of life, just the sort of thing that inspired David Hume's admiration (see chapter 2). Can these adaptations be produced and maintained by natural selection? This process is normally thought to depend on the inheritance of those features that aided, directly or indirectly, the reproduction of ancestors. Worker ants or bees do not reproduce, and no worker can inherit an adaptation from any ancestral worker. This logical difficulty so impressed Darwin that he discussed the possibility that the adaptations of sterile workers must "at once annihilate the theory."

Obviously, if natural selection depends on the survival of the fittest for reproduction, it cannot explain the adaptations of worker ants or bees. But suppose the process can be based on fitness for genetic success rather than reproduction. A worker does not reproduce, but her genes are present in varying proportions in her relatives. If those relatives are successful in reproducing, they are passing her genes to future generations. This argument applies with special force to the kind of life history I am discussing. If a worker were to produce a son or daughter, half her genes would go to each offspring. If her mother produces a daughter (the worker's sister), three-quarters of the worker's genes will go to that new female.

It is important to understand why. The worker's father, who fertilized the queen, had only one set of genes, because he came from an unfertilized egg. All his offspring get the same genes and are like identical twins in these genes from their father. Only the mother, who has two sets of genes from which she puts a randomized single set into each egg, provides any genetic variation. Hymenopteran sisters therefore have a three-quarter relation to one another, but only one-half to each offspring, if they produce any. For a worker, there is more payoff (genetic success) in her mother's reproduction than in her own. So a pedigree diagram for this group of insects looks a bit different from the previous one. The three-quarters genetic similarity between sisters is balanced by the one-quarter relation of a brother to a sister. He shares none of the genes that the sister got from her father, because the brother has no father.

If this seems confusing, just remember this basic definition of genealogical relationship. Point your finger at some random gene on some chromosome in one individual and ask: What is the probability that this gene went from the same source to some other individual? This is the relation-

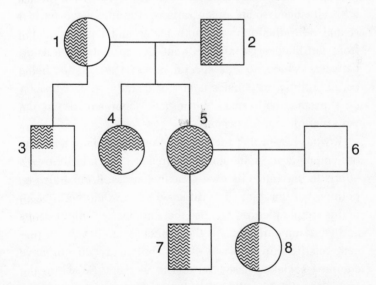

Pedigree diagram showing degrees of relationship for female bee or ant (#5) with various other individuals. Note especially the difference in relationship of a sister (#4) and a brother (#3).

ship of the other individual to the first. Such relationships can be asymmetrical in the Hymenoptera. A randomly selected gene in a female has a 25 percent probability of being in her brother, and this makes him 25 percent related to her. A randomly selected gene from the single set in a male has a 50 percent probability of being in his sister, so that she is 50 percent related to him. The especially close relationship between sisters is often used to explain why, in this insect order, females so regularly forgo their own reproduction and devote themselves to their mother's.

It has been pointed out, in confirmation of this theory, that advanced social organizations have evolved in this order of insects about a dozen times independently, and in every case the societies consist solely of females. Males (drones) make no contributions to the economy of the hive. They leave shortly after they emerge from their brood cells. The importance of the three-quarter relationship is confirmed by comparison with termites, another order of insects with advanced social organizations. Termites have normal sexual reproduction, with both males and females arising from fertilized eggs, and without the special relationship between sisters. So this special reason for societies being based entirely on females is absent in the termites. Exactly as expected, both sexes participate about equally in the workings of termite societies.

It would seem that the elaborate organization of a honeybee colony is an incidental consequence of each individual's efforts to maximize its own genetic success. It must also be pointed out that biologists differ widely in their acceptance of this simple picture, because there are many complications and worrisome details. All the Hymenoptera have a 75 percent relationship between sisters, but many do not have complex societies based on sterile workers. So while the special genetic relationships in this order of insects may make the evolution of complex societies more likely, they are neither necessary nor sufficient. The mothers in nonsocial species raise their own young with complex nurturing behaviors and no help from males. If they were to evolve elaborate societies, would you not expect them to be based on females? The exclusion of males from the social life of the colony would simply be an extension of their ancestral exclusion from parental roles. In many social species, workers may only be half sisters, because the queen mated with more than one male. Some of her daughters may have the same father, some different fathers. A complication to this complication is that daughters of multiply mated queens produced at the same time may be mainly from the same father.

One of the great delights of scholarly pursuits such as biology is that we can all form our own opinions on any issue. There is clear consensus about many important questions, and it is wise to follow it when there is no reason to challenge it, but the jury is still out on just how socially important the 75 percent relationship between full sisters is in the Hymenoptera. But there is a strong consensus on the most basic issue: each individual social insect, no matter how large and complexly organized its society may be, can be regarded as maximally devoted to getting its own genes into future generations. It pursues this goal by adaptive choice of options open to it.

The tactics required for a honeybee's genetic success and that of a typical mammal are radically different. A mammal's adaptations mainly serve its individual interests through that individual's reproduction. A honeybee's adaptations are mainly individual features that serve that individual's genetic interests through its mother's reproduction, which depends on the survival and reproduction of the colony. Such quantitative measures as the amount of time a bee forages for nectar would be optimized in relation to the interests of the colony, not the forager. Many observations indicate that this is so, and give the consistent impression that the colony is a tightly organized team with all members devoted exclusively to group interests. The intricacies of the adaptive organization of the social insects have suggested, to many biologists, the term *superorganism* for entities such as honeybee colonies.

THE SCARCITY OF INDIVIDUAL SUBORDINATION TO GROUP INTERESTS
......

The same kind of analysis should be used to identify the purpose of the adaptations of any other organism or group of organisms. There are natural-history accounts for general audiences that tell us that a salmon leaves its rich hunting grounds in the ocean and migrates up rivers to a remote mountain stream in order to propagate its kind, even though that surely means its own death. Is that what a spawning salmon is doing, sacrificing its life for the continuation of its species? Its spawning may perpetuate its genes, but surely it also perpetuates its species. Does it matter which we think of as its real purpose?

It does indeed, because an examination of the details of its effort supports one conclusion and refutes the other. When a female spawns, she puts her own eggs safely under a layer of gravel in some spot chosen for its special suitability for salmon egg development. In so doing, she may be dislodging previously laid eggs, which then have very little chance of survival. If she were to avoid a previously used area and settle for a slightly less favorable location, the total productivity of young salmon would be increased, even if some other female would have a bit more genetic success than she would. Every action by a spawning female salmon is just what we would expect for an effort to maximize her own genetic success, regardless of its effect on group productivity.

The males are even more obviously competitive. They fight ferociously with each other for the opportunity to fertilize eggs laid by females. One male is adequate to fertilize almost all the eggs of many females, so why the violent and exhausting effort? Because the whole point of the reproductive behavior of a male salmon is to win for itself the greatest attainable proportion of the fertilizations of eggs produced by the local females. Everything it does is obviously a part of this effort to do better than competing males. Nothing it does suggests that it is trying to maximize the number of young salmon produced, or their collective rate of survival, or any other result that might be seen to serve the general welfare of its population.

Another comparison is particularly apt in identifying what Aristotle's "complex action" must really be, even if Aristotle missed it. What happens when (1) an independent individual, plant or animal, is threatened with death, (2) when a honeybee colony is threatened with destruction, and (3) when a population is threatened with extirpation? The answers to the first two questions are the same: the individual or the colony takes emergency measures. An animal fights back, or flees, or hides in a retreat. A plant's response is less obvious, but a plant suddenly attacked by a large number of chewing insects will often change its metabolism so as to devote resources less to growth, more to making defensive toxins.

The bee colony likewise fights back, and may do so in ways that make its functional unity particularly clear. Individual bees may not only risk their lives but actively sacrifice them for the good of the colony. When a bee stings, the sting may come out and take on an existence of its own, actively pumping venom into the stung animal even if the bee itself is brushed away. The breaking loose of the sting always kills the bee, but this in no way reduces the enthusiasm of its attack. It thus clearly shows its priorities: my life is worth very little; it is the survival of my colony that matters, because only thus can my genes survive.

What does a population do when threatened with extirpation, for instance, the reduction of the sockeye salmon stock of the Yukon River to 1 percent of its normal size? Nothing special happens at all. The individual salmon keep on with their normal activities, each trying to reproduce more than its neighbors, with no regard to effects on the stock as a whole. Individual salmon respond to individual threats in adaptive ways, but salmon populations take no concerted action to avoid being wiped out. Their populations show no functional organization like that of a bee colony.

FOR THE HARM OF THE SPECIES
......

The absence of a functional organization for populations or species actually has worse consequences than might be imagined, because natural selection within the groups can produce results that not only fail to help the group as a whole but may be harmful, or at least systematically wasteful. One example is the effect of selection on a population sex ratio. The really basic question of why reproduction is often sexual is explored in chapter 5. Here I will just assume that it is, and is made up of males and females rather than hermaphrodites. What then determines what fraction of the total is male and what fraction female?

The history of thought on this question is rather curious. Darwin worried about it a little, but shrugged it off. It was then totally ignored until 1930, when Ronald A. Fisher, one of the patriarchs of modern Darwinism, tersely provided the essence of the currently accepted idea. This explanation invokes what is now known as frequency-dependent selection, a rather elementary idea in traditional economic reasoning. Suppose you are equally skilled as a smith and a carpenter, and have the resources to set up shop in a village for one trade but not both. You know that the village already has a smith. Should this influence your decision? A well-known example of the same principle, applied in fact to sex ratio, is provided by Shakespeare in *The Taming of the Shrew*. Think of the trouble Baptista would have saved himself if he had fathered a son and a daughter instead of Bianca and Katherina. Minimizing your children's competition for mates is a good idea if you want to maximize your production of grandchildren.

Problems in frequency-dependent selection in biology are often analyzed by what is known in game theory as a payoff matrix, and sex ratio provides the simplest possible example. In the illustration, the individual labeled *player* has a choice of being a male or a female. The next individual it encounters, labeled *opponent*, will be one or the other. Whichever our player chooses to be, if the opponent turns out to be of the opposite sex, it wins in its game of reproduction. The winning is represented by the score of *1* in the matrix. If the opponent is of the same sex, there is no reproduction, no winning, score *0*.

		opponent	
		male	female
player	male	0	1
	female	1	0

Payoff matrix for the sex-ratio game

What advice do you give the player in this kind of game? Unless you know something about the sex of the opponent, there can be nothing wiser than flipping a coin. But suppose you know that the males are slightly in the majority in the population. Now you can offer sound advice: be a female; or, if it is too late for that, have a daughter, not a son. A closer-to-home example: you are a lecherous heterosexual man and there are two singles' bars in town. One of them has mostly men in it, the other mostly women. Which will you choose?

Selection on sex ratio operates by favoring any individual who becomes a member of the minority sex or produces mostly that sex among its offspring. The expected immediate result of this selection is to increase the abundance of the minority sex. The long-term result is that the minority sex stops being in the minority and the sex ratio stabilizes at equal numbers. An example is found in the approximately equal numbers of men and women in the world. As long as this condition prevails, men will, collectively and on the average, produce about the same number of babies as women do. Neither sex has an advantage, and selection on sex ratio disappears. It will reappear and act to reestablish the nearly equal numbers if ever this equilibrium is disturbed.

All sorts of questions will crop up at this point: At what age do we expect equal numbers of males and females? What is the effect of sons and daughters having different mortality rates or different costs to the parents? What if males and females mature at different ages? Why is there, in fact, a slight preponderance of boys at birth? These are questions that biologists have discussed at great length, but their answers imply quite minor quantitative modifications of the 50:50 ratio expected from simple frequency-dependent selection. This selection always favors the minority sex. This is true regardless of how it affects the well-being of the group as a whole.

And the effect can be decidedly negative. Richard Dawkins, in his book *River Out of Eden,* discusses the dramatic example of elephant seals. In the breeding season, the females come ashore on suitable beaches to give birth and nurse their babies and be fertilized for next year's births. The beaches are crowded by adults, a scattering of individual males each with a large harem of females. This is not the adult sex ratio; it is merely the sex ratio of reproducing adults. The true ratio is not far from equal. This means that most of the males are unsuccessful. For every male with a harem, there are many celibate bachelors. They represent a waste of resources, because only a small fraction of them will reproduce. Yet because of frequency-dependent selection, the population goes on, generation after generation, producing about the same number of males as females.

Actually it is worse than would be inferred from just the equality of numbers. There is a gross difference in size, the males being far larger than the females. This is because only the biggest and strongest males have any hope of winning mates against the fierce competition from other males. So frequency-dependent selection keeps the population producing a wastefully large number of males, and sexual selection goes on making each male wastefully large. A successful male, over his lifetime, consumes far more food than does a successful female, who bears the entire physiological burden of reproduction for both her own and her mate's genes. Our own species is afflicted with the same difficulties, but fortunately not to the same extent. The recent feminist slogan "men are not cost-effective" is entirely correct biologically: there are too many of them; they are too big; they accomplish less than women per unit of resources consumed.

The formal game of prisoners' dilemma, for which traders' dilemma would be more apt a name, has a payoff matrix rather different in form from the one in the previous illustration, but shows another way in which natural selection may have negative effects at the group level. Suppose some evening you are driving a car with a German license plate through a little town in Switzerland to your home in Italy. You notice a store that sells computer supplies, and you remember that you need diskettes. You stop because the store is selling for ten marks what would cost you twenty at home. So you are willing to pay the ten marks and be a winner by ten marks, according to your evaluation of the goods. But wait—you might do even better. You happen to have some worthless counterfeit marks. It is too dark in the store now to notice, and the storekeeper won't see until morning that the money is not real. By then you will be in another country, and it is unlikely that you will ever be back here again. Your situation is that described in the matrix. H represents the strategy of being honest, and paying real money, D the dishonest use of the counterfeit. What should you do? Obviously, if self-interest is your only motivation, you should be dishonest. That way you get the twenty marks' worth of goods for nothing. If you paid real money, you would give up ten marks and have a net gain of only ten.

		opponent	
		H	D
	H	10	−10
player	D	20	0

Payoff matrix for the trader's dilemma

Unfortunately, you are not the only player with a dishonest option. The storekeeper has diskettes that he knows to be grossly defective. No buyer will realize this before getting them home and trying to use them. The storekeeper notices your German car and your Friulano accent and thinks, "I'll never meet this guy again. Why waste good merchandise when I can foist off this worthless box?" What should he do? Again, self-interest provides one clear answer: be dishonest! The net expectation from all this rational decision making should be clear from the payoff matrix. If everyone in such a game were consistently honest, everyone would win ten marks per game. But what happens in such an honest society if a cheater appears? The cheater's winnings are greater, and some honest player is penalized. What happens if all players cheat? No one wins. The storekeeper gets worthless counterfeit money, the traveler some worthless diskettes. Yet dishonesty remains the best policy for everyone, because of a simple rule apparent from the payoffs. *No matter what your opponent does, you do better by cheating.* So natural selection proceeds to establish this rule, and reduces the payoff for everyone to zero from a potential positive gain.

I will mention only one of many possible biological applications of the traders' dilemma. There is often an optimal group size for its members, for instance, the number of fish in a school in a pond. When a predator attacks, it is likely that no more than one fish will die, because, once one is caught, the others can get away. This means that if there are m fish in the school, the probability is $1/m$ that a given fish will be the next victim. Obviously the safest place to be is in the largest school available. Unfortunately, the bigger the school the greater the competition for food and the less there will be for each fish. The optimum school size will be that with the greatest excess of benefit, from predator avoidance, over cost, from decreased nutrition.

Suppose ten is the optimum number, and a fish finds itself in a school of twenty. What should it do, purely from the standpoint of self-interest? If it stays, it and all the others will suffer from a food shortage. If it leaves, it would help all the others by increasing their nutrition, but it would expose itself to much greater risk of death from predation. It could well be that, from the standpoint of its long-term fitness considerations, it is better to keep its risk at $1/20$ for the next predator attack and make do with a deficient diet. If so, the twenty fish will continue to swim together, even though, from every individual's perspective, it would be better to break up into two schools of ten.

A human level of rationality in this situation might well result in the two optimal groups. An individual could assume the lead and say, "Look, fellas, there are too many of us. Let's all us on the left side of the school turn left, and you guys on the right head the other way. Then we will achieve the optimum trade-off between predation hazard and competition for food." Unfortunately, the only decision making the fish can manage is a simple "I should stay in this bad situation" versus "I should move away into a worse situation." This inevitably, in the dilemma described, results in schools that are too big, and at least one study suggests that this happens regularly in nature. An enormous number of other examples could be described that would illustrate the principle that, although some groups, such as honeybee colonies, are functionally organized, most animal groupings are not. They are just mobs of self-seeking individuals. In the next chapter I return to the examination of evolved mechanisms, with emphasis on two related questions: How are they produced (development), and how do they work (physiology)? I also consider the more fundamental question of how such problems are legitimately resolved.

CHAPTER 4

···

THE ADAPTIVE BODY

The genotype of an organism is commonly likened to a blue-print for a house, but this analogy is "dreadfully misleading," as pointed out by Richard Dawkins, the justly influential author of *The Selfish Gene*. As he notes, you can point to part of a house, like a north window, and then to an exactly corresponding rectangle on the blueprint. There is no such correspondence between a part of an organism and a part of its genotype. The best that might be done is to point to a location on a chromosome and note that the gene at that position specifies the structure of a protein, a part of an organism that may be present in an immense number of copies in an immense number of cells scattered in many different tissues. And even this sort of correspondence is limited. Some proteins are assembled from components made by more than one gene.

Dawkins likewise notes that you can easily use a house as the model for making a blueprint that can then serve in the building of a similar house. To specify a genotype from a knowledge of some advanced developmental stage of an entire organism would be not merely difficult but quite impossible from any current understanding of developmental genetics. Dawkins suggests the relationship between a recipe and a cake as a better analogy. The recipe is a set of directions for producing the cake, but you can seldom identify a statement in the recipe that corresponds to a specific part of the cake. Likewise, it is much easier to make the cake by following the recipe than to infer the recipe from examining the cake.

Genotypes differ from both blueprints and recipes in always containing instructions for getting instructions, rather like modules in an interactive computer program. Genotypes are full of statements such as, "If x is less than k, then do y, or else do Y," and "Note x, then make $y = f(x)$." For example, if you get more than a certain amount of royal jelly, develop into a queen, otherwise become a worker (for a larval honeybee); or, note the level of incident sunlight, and make a directly related amount of melanin pigment (for a human skin cell). It should, of course, be realized that a genotype's instructions for an organism are far more complex than a blueprint's specifications for a house or a recipe's directions for a cake.

But genotypes, blueprints, programs, and recipes all alleviate the complexity by making use of whatever simplifications might be available. In planning a house, you need not give detailed instructions for the laying of every brick, even though the laying of even one brick may require considerable skill exercised by an extremely well trained bricklayer. Building a house from a blueprint may require preexisting bricks and bricklayers, standard electrical fixtures, trained electricians, and much more. Likewise, the use of genetic instructions in development may assume the prior existence of a long list of available components and reliable processes. A human genotype assumes the immediate availability of various carbohydrates, amino acids (the modular components of proteins), and other essential molecules. Initially it gets them from the maternal bloodstream, later from milk, later still from a variety of ingested foods.

A human genotype also assumes the reliability of the

laws of physics and chemistry. If a gene makes an enzyme molecule useful in splitting a protein into component amino acids, the same gene can do that job many times in many different places. If available molecules spontaneously combine in certain ways that happen to prove useful, this chemical process will be exploited wherever it might help. It might even be said that organisms, wherever possible, delegate jobs to useful spontaneous processes, much as a builder may temporarily let gravity hold things in place and let the wind disperse paint fumes.

This being said, it remains true that the development of any organism, for instance, a one-celled paramecium, is an unimaginably complicated process that requires a huge array of precise genetic instructions. We have considerable understanding of this developmental program at a basic molecular level. We understand how DNA is activated and transcribes its information to RNA. We understand how RNA specifies protein structure. We understand how protein molecules fold into certain shapes and interact with other such molecules to form filaments or sheets or other simple structures. We do not know much beyond that. We know very little, for example, about how the protein-based machinery of the cells interacts with calcium ions and other nutrients to make a left collarbone in exactly the proper size and shape and position in the human body. We can, of course, establish cause-effect relations in the making of a collarbone. We can show that no adequate collarbone, or any other bone, will be made in the absence of various minerals, Vitamin D, and other substances in minute traces. We know that altering hormone concentrations can alter the growth of certain bones, like those with different sizes or shapes in the two sexes. We can show that usage can affect a bone's size and shape and distribution of density, and that such effects of usage are normally adaptive. Moderately increased stresses on one part of a bone will often strengthen that part.

Yet not even a complete list of such cause-effect relations

really adds up to an understanding of development. The distinguished British biologist John Maynard Smith has expressed a characteristically simple but profound comment on this lack of understanding:

> One reason why we find it so hard to understand the development of form may be that we do not make machines that develop: often, we understand biological phenomena only when we have invented machines with similar properties. The shapes of the things we make are, as a general rule, imposed on them from the outside: we do not make "embryo" machines which acquire complex shapes by intrinsic processes.

But why do we not make "machines that develop"? I suggest that it is because development, as found in biology, is far more difficult than any sort of artifice we commonly undertake. We are not smart enough to make machines that develop. The making of a hammer embryo that can develop into a hammer, merely by providing it with access to wood and iron, is far beyond our capabilities—and a hammer is an extremely simple machine.

The "intrinsic processes" of development so impressed the pioneers of experimental embryology that many thought it obviously required supernatural guidance. They found much complex machinery in an embryo, but it was inconceivable to them that such machinery could guide itself. There must be some immaterial, purposive entity in control of the visible machinery. They assumed that this entity in a frog embryo knows how to make a frog and desires to do so. They imagined it guiding the machinery so as to produce that result, even if external forces disrupt the process to a limited extent. This purposive entity made the visible embryonic machinery repair damage, compensate for losses, and head unerringly toward the production of a frog.

This mode of thought is now banished from embryology.

Researchers no longer invoke immaterial entities for the control of development. As I pointed out, their success in finding purely material explanations is quite limited, but their faith in materialistic embryology seems unshaken. I suspect that this faith has been aided, in recent years, by technological successes in artificial self-regulation. We can make thermostatically regulated boxes that attain and maintain a specified temperature, despite heating or cooling from the outside. Modern scientists' familiarity with many such examples of machinery that controls itself no doubt makes it easier for them to believe that the machinery seen in an embryo also controls itself.

VITALISM AND MECHANISM
·······

Supernatural agencies are banished not only from explanations of the development of biological mechanisms but also from modern attempts to understand their operation. When I used the term *protein-based machinery*, I implied that the workings of the human body can be understood in ways closely similar to our understanding of metal and plastic machinery. I assumed that we can, metaphorically at least, take the body apart and see what makes it tick. This understanding can make use of any and all known physical principles, but no modern physiologists would invoke anything immaterial or supernatural. If they fail in their efforts to understand how a part works, it must be because they did something wrong, and perhaps it will be correctable in the next attempt. Biologists never conclude that physics and chemistry are not a sufficient basis for understanding how a biological mechanism works. This banishment of the supernatural from biological explanation is traditionally known as the doctrine of *mechanism*.

The alternative is *vitalism*, which assumes that the workings of an organism, human or other, require something more than physical machinery operating according to the laws of physics and chemistry. The machinery is demonstrably there, but vitalists believe that its autonomy is limited. They envision some immaterial entity operating the machinery, just as a ship, however complex and automated, may still need the mind of a captain to make decisions and initiate control.

In the early years of this century, many professional biologists were unabashed vitalists. But modern studies of the development and workings of the body almost always proceed from the assumption that everything can be physically manipulated and physically understood. I say "almost" because I detect some minor exceptions in the field of neurophysiology and behavior. There are those who believe that the material controls in the nervous system are subject to manipulation by immaterial mental processes. They presume that to understand an animal, it is necessary to investigate not only its brain and other material organs but also its mind.

How is an animal's mind to be studied? Is there a way to ask an animal "What are you thinking?" If a dog repeatedly investigates its empty water bowl, you might say, with some justification, that it tells you it is thinking about water. You might say something similar about a computer when your prolonged idleness at the keyboard results in its screen-save behavior; it thinks you are at fault for abusing its phosphor and wasting power. A more impressive kind of answer to the question "What are you thinking?" would be one expressed in a symbolic language. Computers obviously provide such responses. Mine just now expressed its frustration, in unmistakable English, at my failure to put a diskette in a drive that I asked it to activate.

Can animals ever use words? Perhaps so, if some of the ape-language specialists are right. Just as you can ask a deaf student a question and get a verbal answer using sign language, you could ask Nim Chimpsky what he plans to do and get a sign-language response. Is either of these processes really different from a vocal conversation with a person with normal hearing? If Nim is really conveying symbolic meaning with his gestures, the answer is surely no. I have my doubts about apes' abilities to use symbolism, but for a biologist it really doesn't matter. Using a language, whether signed or spoken or printed, is a kind of behavior, like walking or eating or courting. If walking can be understood mechanistically, why not speaking? Both require nerve impulses and muscle contractions, just the sorts of processes mechanism has shown itself consistently adequate to understand.

Most people think a computer can be understood in a purely mechanistic way, even though it can respond to verbal questions with verbal answers, or can initiate a conversation by asking the questions itself. Workers in artificial intelligence have come a long way toward programming conversational skills into computers, so that it is increasingly difficult to decide, as you input questions and responses on a keyboard, whether you are communicating with a real person behind the wall or merely the switches and chips in front of you.

Yet even though it is increasingly difficult to distinguish artificial from human intelligence, it is still rather easy. My own reliable way of deciding whether I am dealing with artificial intelligence or a real person responding to my keyboard input would be to look for a sense of humor. I don't mean the ability to recite a joke; a computer can easily be programmed to do that. I mean the ability to respond with a really funny one-liner related to what I just said. I do not expect in my lifetime to interact with a switches-and-chips equivalent of S. J. Perelman.

Ultimately the only mind you can really know is your own, not those of computers or animals or friends. I think that this removes the domain of the mental from biology and from material science in general. Descartes, wisely, said

"Cogito ergo sum," not "Cogitamus ergo summus." In this book I assume the adequacy of mechanism and the impropriety of vitalism in any form for all phenomena in biology. (I have more to say on mentalism in discussing brain function later in this chapter and again in chapter 9.)

THE MACHINERY OF THE BODY
······

The title for this section is that of a deservedly influential textbook of physiology by Anton J. Carlson and Victor Johnson, often used in college biology courses in the middle years of this century. It is a title that surely implies the adequacy of mechanism in answering questions about the operations of an organism, whether human or other. There are many possible illustrations of the operation of the machinery of the body. Here I will start by reconsidering one discussed in chapter 1, the manipulative usefulness of the human hand.

Manipulation, like any other motor activity, requires muscle contraction. The manipulation muscles are mostly in the forearm. As you alternately squeeze and release the remaining pages of this book with your right hand, you can use your left to feel the muscle contractions and relaxations. You can also feel the tensing and slackening of tendons in your wrist and palm. These tendons are the cables that convey the force from the forearm muscles to the bones of the fingers. They pass through precisely aligned systems of tubes and grooves to accomplish this, and each tendon ends with an attachment at exactly the right spot on the right bone to produce the needed squeeze or straightening of the finger. There are many such tendons, from many separable bundles of muscle, so that the fingers can perform many kinds of motion, such as curling up and straightening out and waving from side to side. There is some sharing of muscle groups between fingers, so that we are limited in our ability to move the fifth digit without moving the fourth.

All this manipulative ability is understandable in detail as a purely mechanical consequence, via tendons and their attachments, of the contraction of muscles in the forearm (and at the base of the thumb). A finger does not curl up or straighten out because it wants to, but because it is forced to by the pulling of tendons. These tendons pull only because they are pulled by muscles. Why do the muscles pull? Their pulling is an active shortening of the muscle as a whole, and this shortening in the realm of the readily visible reflects a shortening process in the realm of the ultramicroscopic. Muscles are densely loaded with parallel protein fibers running from one end to the other. These fibers actively shorten, and provide the muscle with its forceful contraction, because they forcefully fold, somewhat like the bellows of a camera. Their normal relaxed state is stretched out. Their contraction requires a supply of energy from a substance called adenosine triphosphate (*ATP*). The energy taken from the metabolism of this substance comes ultimately from the food we eat and the oxygen we breathe. Understanding all this requires an impressive array of technical knowledge of the molecular machinery of the cell, but this knowledge is entirely of physics and chemistry. No subtle immaterial processes are employed.

A muscle's protein fibers fold and make the muscle contract and pull on a tendon because they are stimulated to do so by a nerve impulse. People may think they understand the conduction of electrical impulses by nerves because they think they understand the conduction of messages by telephone wires. What goes on in a telephone wire is, in fact, more complicated than most people realize. What goes on in a nerve fiber is incomparably more so. There is a rough analogy because both processes seem extremely rapid, both sorts of impulses travel through a long, thin conductor, and both involve electric charges. They have very little else in common, and calling the nerve impulse electrical can be misleading. There is no electric current moving along a nerve. A nerve impulse is a wave of exchanges of electric charges between atoms of the outer and inner surfaces of the membrane that bounds the nerve fiber. The wave typically travels many meters per second, but this speed is a minute fraction of that of an electrical impulse in a wire.

BRAIN AS MACHINE
......

I will bypass the physics and chemistry of nerve impulse conduction, as I did that of muscle contraction, and go on to the next question. What makes the nerve conduct an impulse to the muscle? For any action we would speak of as voluntary, such as turning a page, it is always because of something that happens in the brain, the machinery behind everything voluntary, and behind much else besides. Unfortunately, the brain does not lend itself to understanding as the machinery of motivation or thought in the way that a nerve is understandable as a signal transmitter, or a muscle as a producer of mechanical action, or a tendon as a transmitter of such action.

A human brain or rat brain or snail brain has many parts and looks complicated. The parts are connected to each other by nerve fibers. Each part, if examined closely, is found to have connected subparts. Subparts examined with magnification show still more complicated detail, and the pattern continues down to the complicated parts of the nerve cells themselves. In this way we can amass an enormous amount of knowledge about a brain's structure, but the knowledge does not lead to much functional understanding. We do not understand the brain as a machine for thinking in the way we understand the heart as a machine for pumping. We also have a great wealth of information on brain biophysics and biochemistry, such things as alpha waves and brain hormone secretion. Again, the wealth of information supports a poverty of understanding.

Of course, we do have some comprehension of the brain's control of various kinds of behavior and cognitive capability. Injury to some parts of the brain interferes with vision, injuries elsewhere interfere with speech or short-term recall. It is not surprising that an injury where the optic nerve joins the brain would interfere with vision, and it is not surprising that injury to the part it joins might have a similar effect. But we do not really understand why. Brain injuries remote from the optic nerve may also result in visual impairment. The other part is no doubt connected to the area where the optic nerve enters, but all sections of the brain are connected.

The sheer complexity of the brain, in relation to a function such as vision, is no doubt a factor in making vision and the processing of visual information difficult to understand. The problem grows even more serious when we turn from sensory mechanisms to what we are really interested in here, like the decision to turn a page. I suggest that the difficulty arises from two factors. One is that, for most of the workings of the body, big effects often require big causes. A violent kick at a football requires a large expenditure of energy by large muscles. The decision about whether to kick it, and how hard and in what direction, may depend on physically trivial interactions between microscopic parts at many places in the brain. The processing of information, even for violent thoughts, does not require the energy needed for violent action.

The second source of difficulty is even more serious. What is it about the heart that makes it so understandable as a pumping machine? I suggest that it is simply the fact that a description of the heart's physical attributes and a description of the pumping make use of the same terms with the same meanings. The terms are all those of ordinary physical science: mass, velocity, pressure, length, and so on. We understand why the mitral valve has to be oriented in a certain way for effective pumping into the left ventricle. The mitral valve is a material entity with physically measurable dimensions and mechanical properties, and pumping is a material process describable in the same terms.

By contrast, the material descriptors of the brain—size, shape, density, charge, and so on—say nothing about thought, always described by a different set of terms: desire, planning, analysis, recollection. The physical and mental domains lack shared descriptors, and we cannot start reasoning with concepts applicable to one domain and reach a conclusion applicable to the other. No reasoning from the acidity of a brain hormone or the mass of the olfactory lobe can explain why a friendship went sour or measure a burden of grief.

My inclination is to purge all biological discussion of mentalist interpretation. If I should propose that a mosquito turns upwind whenever it detects increased carbon dioxide so that it can find a breathing animal to feed on, I am talking about its adaptive programming, not about its understanding or thinking. Likewise, when I propose that Suleiman the Bloodthirsty, a Moroccan potentate some centuries back, amassed a large harem in order to maximize his genetic representation in future generations, I would not be implying that this is what he consciously wished. I would be talking about the adaptive programming that was precisely organized for this effect. As a compulsive lecher, he might even have been mentally disappointed at genetic good news (for example, that his favorite sex partner was pregnant). I have more to say in chapter 9 about the folly of trying to make reasoned connections between domains that lack shared descriptors.

THE MOLECULAR MACHINERY OF THE BODY
·······

A mere part of a chapter on the body's adaptive mechanisms, even if it is only one kind of body (human), can be no more than a molehill beside the great mountain of physiological lore that has accumulated. My goal is to try to convey some feeling for the consummate effectiveness of that machinery.

It runs so well that we are surprised when it fails in even a minor way. It should be just the opposite. To survive for even a few minutes is a magnificent accomplishment. If you survive for more than a few days, it is a marvel far greater than any of the miracles of mythology. My favorite statement of this idea is from the marine biologist George Liles: "the cells and organs that make life possible had better be well designed, because the job of living is formidable." He was led to this remark by considering the machinery that routes water through the feeding sieve of a mussel. This is a simple job compared to the avoidance of cancer, a problem I examine in chapter 8. I will end this chapter with two rather different examples of the body's machinery, two among the endless number that I could have used.

Consider an actively working human cell. Any cell will do—nerve, white blood corpuscle, or skin cell—and we need not worry about what the cell as a whole accomplishes. I will concentrate on just one detail of how it looks after itself so that it can go on with its normal contribution to the whole. Whatever it does, and whatever sort of machinery it uses, it will require energy. It gets this energy by processing nutrient substances, and there must be elaborate machinery for delivering these substances to the cell and for transporting them to key positions within the cell. I will skip all that and deal with only one kind of nutrient.

The nutrient, the sugar glucose, is delivered to a structure called a mitochondrion, of which there are often hundreds in the cell. Each mitochondrion is a complex structure, with a smooth outer membrane and an inner membrane with folds projecting into the interior matrix. The mitochondrion is so small that this structural detail is seen only with magnifications of many thousandfold by an electron microscope. Despite its small size, it is immensely complex, with many kinds of specialized features on the membranes. With these structures it subjects the glucose to a series of reactions that provide the cell with some of the energy it needs. Each product of each reaction must be delivered to the next appropriate enzyme and mitochondrial structure for processing.

It is tempting to compare this mitochondrial processing of molecules to an industrial assembly line, but the geometry of the processing is not the simple linear arrangement suggested by this analogy. Materials must be shunted repeatedly from the matrix through the inner membrane to the space between the membranes and back again. Finally, the waste product carbon dioxide diffuses away, and some of the energy-rich compound ATP is exported through the outer membrane for use elsewhere in the cell. The rates of many of the individual reactions are precisely controlled by what are termed allosteric feedback loops. An enzyme that mediates one of the reactions may itself be sensitive to the product of that reaction. As the product accumulates, the enzyme reacts by changing its form so that it is less effective in mediating the reaction. Other enzymes speed up in response to an accumulation of the sorts of molecules they process, much as a bank teller may try to work faster as the line of customers lengthens.

This incursion into the world of molecular dimensions gives a new perspective on David Hume's admiration of the "various machines" and "even their most minute parts" (see chapter 2). This admiration was based on the gross examination he could make with his naked eyes. Modern microscopy and biochemical techniques provide a far more persuasive basis for admiration.

There is another reason for considering the world within

the cell in relation to plan and purpose in nature. In dealing with the activities of separate individuals, evolutionary biologists must attend to complex interactions among conflicting interests. Relations between mates, between parent and offspring, between siblings, between neighboring territorial competitors, between host and parasite, are characterized by a complex array of cooperation, conflict, compromise, winners and losers, perhaps stable stalemates. In trying to understand what goes on in a single cell, do we need to invoke any of the theoretical panoply of optimization, kin selection, and payoff matrices? The answer is an uncompromising yes.

Consider that mitochondrion I discussed. Mitochondria are not directly assembled in a cell from components. They always arise by the division of preexisting mitochondria. They have their own genotypes coded in their own DNA. As a general rule, all the mitochondria in a cell have the same genotype, and we expect minimal conflict among them, but the cell itself has a different genotype. Could there be conflict between the cell's nuclear genes and its mitochondria? Yes, there could, in important ways, and I will mention only one obvious example.

The mitochondria are transmitted only in the female line. Those in an offspring are supplied entirely by the egg. This means that the mitochondria in a male are doomed to extinction. It obviously pays a mitochondrion (and its genes) to stay out of males. Those in an egg should use their influence to make the egg reject a male-determining sperm. If fertilization is in fact by a sperm with a Y chromosome that would ordinarily cause male development, a mitochondrion should do everything it can to frustrate the Y effect and turn the individual into a female. The fact that we do not ordinarily see such phenomena does not mean an absence of conflict. It means that in this particular conflict, there is a winner (nuclear genes collectively) and a loser (mitochondria). The winner-loser concept plays a decisive role in modern ideas on the evolutionary origin of maleness and femaleness (chapter 5).

Much of what goes on in a cell is directly controlled by nuclear genes. Could there be conflict among them? Here again the answer is yes, even though each gene depends on all the others for maximizing the fitness of the genotype in which they now find themselves. If reproduction is sexual, that genotype is temporary, and those genes will no longer depend on one another in the future. This future independence can mean conflict now. Suppose a gene has the ability to duplicate itself so that it is present not just once but twice in the genotype. The immediate effect is likely to be deleterious to the collective interest, because that is what we always expect of an arbitrary change in any highly perfected machinery. The duplicator doubles its representation at an unknown cost to all the other genes. This is not a hypothetical idea. The literature of genetics is full of examples of such phenomena, under such terms as tandem duplication and transposable elements. Moreover, genes' actions may be timed differently according to whether they came from the father or the mother (genetic imprinting), if such timing gives special advantages to the paternal or maternal genes in the future. (I discuss this further in the next chapter.)

There is also considerable recent attention to the mechanisms by which the well-behaved majority of the genes attempt to impose conformity on the occasional rogues. The majority must somehow enforce limits to gene duplication, mitochondrion reproduction, and other attempts by the cell's components to achieve short-term gain at long-term cost. Disruptive reproduction of whole cells (cancer) is another kind of rebellion that must be suppressed (chapter 8). The detailed workings of a cell are inevitably a complex system of compromises, controls, and safeguards understandable only from a Darwinian perspective.

In the formation of eggs and sperm, there is a special opportunity for a gene to enhance its own fitness at the expense of its neighbors'. The processes that produce these reproductive cells are supposed to conduct an absolutely unbiased game of chance. If an individual got gene A from its mother and A' from its father, it should produce A-bearing and A'-bearing sperm or eggs in exactly equal numbers. But suppose A' is a mutant that somehow biases the process so that more of it is passed on than of A. This could greatly enhance the fitness of A' at the expense of A. It could also be at the expense of all other genes in the population, because it would mean that A' will be increasing, no matter what effects it may have on the genetic success of its bearers. There are a number of examples of this sort of conflict in nature. A mouse gene can maintain an appreciable frequency in wild populations by biasing sperm-production in its own favor, even though it greatly decreases the fitness of its bearers.

Attention to the ways in which natural selection shapes biological phenomena has traditionally been more of a preoccupation of ecologists and animal behavior researchers than of physiologists and molecular biologists. Today the importance of evolutionary insights is being more widely recognized among biologists who study the inner workings of cells. An attempt to understand cellular phenomena entirely on the basis of the chemistry and physics of molecular interactions, without any consideration of adaptive significance, is about as sensible as trying to understand the Napoleonic wars entirely from a knowledge of ballistics.

This and the previous chapter dealt with the essentials of the process of natural selection, how it operates, on what it operates, and what it can and cannot accomplish. I will now turn to a more detailed look at what it has accomplished in the machinery that produces and operates a living organism, with main emphasis on the human organism. I will also be concerned with the scientific methodology that has proved useful in providing an understanding of how the evolved machinery operates.

CHAPTER 5

The title of this chapter is that of a pioneering 1971 article by the British biologist John Maynard Smith. His works, especially his book *The Evolution of Sex,* are classic contributions to our understanding of the topic, and form the basis for much of this chapter.

THE ORIGIN OF SEXUALITY

In this chapter, I will use of a number of convenient oversimplifications: there are three main groups of organisms—bacteria, plants, and animals; all plants and animals reproduce sexually, at least occasionally, but may also have asexual reproductive processes; bacteria reproduce only asexually. This last claim does not mean that a bacterial cell always transmits its own genotype to descendant cells. On rare occasions different cells may exchange one or more genes, and, of course, even within a lineage the rare event of mutation may occur. It remains true that the normal bacterial reproductive process is strictly asexual. One cell divides into two, each of which gets the original cell's genotype intact. Nothing observable in bacteria is known to indicate for sure how sexuality originated in the ancestry of plants and animals.

In fact, very little is known about the origin of sexuality. It must have arisen before the Cambrian period, the earliest part of the geological record that contains abundant fossils of recognizable multicellular animals and plants. The Cambrian began about 600 million years ago. Bacterial life had already been in existence at least two billion years. Sometime in that interval, sexual processes were evolved by the ancestors of the organisms known from the Cambrian.

What the sequence of steps was is difficult to determine, but there is a growing consensus that sexuality arose as a by-product of mechanisms for maintaining the information content of genes. One of several possibilities can be illustrated as follows. Suppose you are proofreading a manuscript and find the sentence *We are no4 prepared for winter*. You know that *no4* cannot be right, because numerals never occur at the ends of words. This is analogous to a gene containing the sequence *CCAXT*. The *X* can not possibly be right, because, whatever it is, it is not one of the expected DNA components. Fortunately, all we need to do here is consult the complementary strand of the DNA. If it is *GGTCA*, we immediately know that the *X* should be replaced by a *G* (*C* always has the paired complement *G*). Manuscript proofreaders must do without this kind of help; no rule tells them whether the *no4* should be changed to *not* or to *now*.

There are times when the genetic bookkeeping is not as easy as in this example. Suppose we find not an *X* but an inappropriate DNA component. Suppose one strand reads *GGTCA* and the other *CCAAT*. The *C* in the first strand needs a *G* in the second, and the second *A* in the second strand implies a *T* in that position in the first. Both cannot be correct; one may be as wrong as a *now* where a *not* should be written, but there is no way of knowing which strand is correct. Wouldn't it be helpful, for either genetic or manuscript proofreading, to be able to consult another copy of the same document? You could read it to see whether to substitute a *C-G* or a *T-A* (or a *not* or a *now*), and then go on with the proofreading.

Even the availability of that other copy is no absolute assurance that you can make an appropriate correction. Suppose the other manuscript says *nod*. It is a real word, but makes no sense in the sentence. What do you do? With no other recourse you can only guess, but this can be hazardous. *We are not ready* and *We are now ready* have quite different meanings, and a mistake could have serious consequences. Fortunately, it is unlikely that the same word in a different typescript would have been misspelled in the same way. Having that other copy would usually be an effective asset for keeping a manuscript true to a writer's intent.

The same reasoning applies to that other kind of typescript, the genetic code. An alteration in the molecular structure of a gene will often be immediately recognizable, by the enzymatic machinery of the cell, as nonsense. If there is another copy of that same gene available, it can be consulted and the nonsense changed to whatever structure the other copy shows. This mechanism will usually keep the genetic message true to its adaptive meaning. Suppose one gene says the equivalent of *not* and the other *now*. Both make sense, but one must be wrong, and there is no way to know which. So the genetic proofreading machinery, while extremely effective, is not perfect. Gene mutations do happen and remain uncorrected, and both the correct and the incorrect form will be passed along in reproduction, with perhaps disastrous consequences for some descendants that get the maladaptive message.

The occasional need to repair genetic damage means that having two copies of each gene available at some stage in the life history could have contributed to the fitness of some primitive bacteria. A mechanism for assuring the availability of two copies could incidentally lead to sexuality. If the two copies at some stage following the proofreading go their separate ways, the genetic processes of segregation and independent assortment, characteristic of sexual reproduction, would be provided.

This genetic proofreading requires that genes from different sources be lined up and compared, but between proofreadings they can go their separate ways. If so, different genomes need to be brought back together again the next time proofreading is needed, but it won't help much to bring together genes from a recent common source. Likewise for deciding what to do about the *no4*, it would do no good to consult a recent photocopy of the same manuscript, which would also read *no4* at the same place. You need an independently derived copy of the troublesome sentence.

So a primitive cell that needs to proofread its genome would be well advised to consult a cell from a different line of descent in the same species. This is what usually happens when a one-celled plant or animal indulges in sexuality. It fuses with a similar cell of another lineage, and the genetic comparisons can take place. This stage of having two genomes in the same cell is typically brief in one-celled organisms. The compound cell divides, with each daughter cell getting one copy of each gene. The needed proofreading has been carried out, and something else may also have happened. Unless one daughter cell gets exactly the genome of one of the original fusing cells and the other that of the other original, some genetic recombination will have taken place. A mechanism for the repair of DNA damage would have incidentally produced the defining feature of sexual reproduction.

But, in fact, none of this seems to have anything to do with reproduction. I proposed two cells coming together to form a zygote with two genomes, which then divides to restore the original two-celled condition. None of the numerical increase of cell number, as implied by the term *reproduction*, has happened, but another quite significant kind of increase can be recognized: if genetic recombination has occurred, two new *genotypes* have been created in the cells formed by the zygote's division. We might have started with a clone of cells of genotype X and another of genotype Y. Then a cell from each clone fused with one from the other, and the resulting zygote divided to produce cells of genotypes W and Z. The number of cells has not changed, but the number of genotypes has. Starting with X and Y, we end up with X, Y, W, and Z.

Safeguarding the reliability of the genetic information in each gene may have been the original function of sexuality, but this need not mean that it is now the only function, or even a currently important one. Both theoretical considerations and an examination of the life histories of modern organisms suggest quite a different significance for this process: it diversifies the offspring, as in the production of the two new genotypes mentioned. This is adaptive in relation to environmental uncertainties faced by the new generation. Such diversification may be maladaptive if an offspring is to develop under conditions closely similar to those of the parent. The next generation would do well to keep what has proved successful under current conditions. Organisms—many plants, for example—that have both sexual and asexual reproduction conform to expectations. If the offspring are to develop immediately near the parent, asexual reproduction is the rule, with structures such as runners or tubers. If the offspring are to be broadcast far and wide into an unknown diversity of new habitats, sexual seed production is what we find.

It is important to identify the sorts of variation and uncertainties that make sexual reproduction, with diversified offspring genotypes, the better strategy. It is clearly not a matter of simple physical variables such as temperature. Temperature and water chemistry are much more variable in fresh waters than in marine habitats, but asexual reproduction is more common in fresh water. It is the biological uncertain-

ties that make sexual reproduction adaptive. A healthy herb-
aceous perennial plant in a weedy field has successfully
coped with a particular array of parasites, insect pests, and
competing plants. In producing offspring that will grow up
close by in the same conditions, it uses its vegetative asexual
process. If it is to produce offspring that must compete with
the unknown exploiters and competitors many meters away,
it will start them off as sexually diversified seeds.

WHY EGGS AND SPERM?
·······

There are microorganisms in which sexual reproduction is
very much like that just described for editing the DNA. Two
similar cells join at a predictable time of day or other cue
and form a zygote, a single cell with two sets of genes. Their
chromosomes line up and perhaps exchange parts, then separate and reconstitute two nuclei, each with the correct
number of chromosomes and genes. When the cell divides,
the daughter cells move apart. Each then grows and may
divide asexually, perhaps for many asexual generations, so
as to produce a clone of many independent cells of the same
genotype, before sexual fusion with other cells again takes
place.

Much the same process may occur in various multicellular algae. An alga may release free-swimming cells, all nearly
identical in size, that fuse with similar cells from another
alga. Each of these compound cells may then divide, but the
clones initiated by the daughter cells stick together and form
a cell cluster that eventually grows into the full-size algal
body. This new algal generation will sooner or later produce
the sex cells (gametes) that initiate the next sexual generation.

This kind of sexual reproduction, with similar-size
gametes fusing to form the zygote, is rare in multicellular
organisms. Usually the fusing cells are extremely dissimilar

in size, one unusually large (an egg), the other quite miniature (a sperm). Why should this be? Why not the situation
first described, with the gametes nearly the same size? Surprisingly, this obvious question was essentially ignored until
1972, when a team of two British biologists, G. A. Parker and
R. R. Baker, and a Nigerian, V. G. F. Smith, proposed a plausible historical account that has largely withstood scrutiny
and new evidence.

Once upon a time, their story goes, the ancestor of multicellular plants and animals made use of equal-size gametes.
A zygote had to have an adequate supply of resources to
endow a new individual, and each gamete carried half the
needed nutrients and other requirements, so that the necessary supplies would be provided. Unfortunately, this situation lent itself to cheating. Perhaps the average individual
made a thousand gametes from each milligram of material
devoted to reproduction. Two such gametes could then fuse
to produce a 2-microgram zygote, which I will assume to be
quite adequate for its normal development. But suppose, in
such a population, a mutation occurred that changed the
gamete-producing machinery so that each milligram of
material is divided into 1,100 gametes of about 0.9 micro-
gram each. They then fuse with normal gametes from other
individuals to form about 1,100 offspring of about 1.9
micrograms. The mutant form has 10 percent more offspring
than the ancestral, and this by itself would mean favorable
selection.

Obviously that cannot be the whole story. Zygotes produced by this mutant are of size 1.9 rather than 2 µg. How
much of a disadvantage is this? The handicap would have to
be more than 5 percent to offset the larger number of such
zygotes. If not, the mutant form would have a net advantage,
with its greater fecundity more than balancing the lower
zygote viability, and the mutation would spread. Soon these
cheater types would be so numerous that an appreciable
number of 0.9-µg gametes would be fusing with each other

to produce 1.8-µg zygotes. How well do they fare? Even if they are sufficiently unfit that 0.9 µg gametes no longer have a net advantage, the situation could stabilize with both the 0.9 and 1.0 gametes maintained in the population. If so, additional evolutionary changes could complicate the picture. If an appreciable number of individuals start life smaller than the original ideal of 2.0 µg, there could be evolutionary changes to increase the ability of zygotes to survive with less than 2.0 mg of material resources. This would ameliorate the disadvantage of the smaller gametes so that they could become still more common.

If a mutation can reduce gamete size from 1.0 to 0.9, why not to 0.8? The possible zygote sizes from the combinations of these three gamete sizes are shown in the table. We should not be surprised to find a wide range in fitness when size ranges from 1.6 to 2.0. Serious mathematical reasoning, applicable to a variety of relationships between offspring size and fitness, is needed to assess the trade-offs, for parental fitness, of different gamete sizes. Those producing the smallest have the largest number of offspring, but of below average fitness. Those with the largest gametes have the fittest offspring, but fewer of them. Those in between have moderate numbers of offspring with intermediate fitness. The effect of selection will depend on the exact values of these quantitative relations.

GAMETE SIZES

		0.8	0.9	1.0
G A M E T E	S I Z E S 0.8	1.6	1.7	1.8
	0.9	1.7	1.8	1.9
	1.0	1.8	1.9	2.0

The mathematical arguments are available in the technical literature for those who want them. Here I will merely summarize by noting that under a wide range of relevant variables, parents with the largest gametes and the smallest do better than those with intermediate sizes. The evolutionary result is that the big gametes evolve to be bigger, and ultimately identifiable as eggs. The small ones evolve to be smaller, and end up as those marvels of miniaturization, sperm cells. The sperm will have to compete with one another for the limited number of eggs available, and will therefore retain the locomotor mechanisms (long propulsive tails in most species) needed in the race for fertilization. The eggs can leave the work of finding a partner to the sperm.

I have described the primeval battle of the sexes, with male winners and female losers, mentioned in chapter 2. When egg-producers reproduce, they must bear the entire nutritional burden of nurturing the offspring. By contrast, the sperm makers reproduce for free. A sperm is not a contribution to the next generation; it is a claim on contributions put into an egg by another individual. Males of most species make no investments in the next generation, but merely compete with one another for the opportunity to exploit investments made by females. Only secondarily, in a few groups of animals, have males evolved ways of helping their offspring get a start in life.

I have proposed *winners* and *losers* in the purely economic sense of providing nutritional resources for offspring. Females are the losers because, if they are to reproduce, they must bear the whole nutritional cost for both themselves and their mates. This need not mean that men have easier or happier lives than women—the opposite is true with respect to the legitimate human values of survival and health. The biology of reproduction forms a basis for ethnic and religious traditions that facilitate the oppression of women by men and of children by both, with everything arranged so that the men end up the biggest losers. In seeking evidence on this conclusion, I would suggest that sex differences in the statistics of morbidity and mortality provide a good start.

In a few species, females have entirely rid themselves of

the male parasitism. If a female makes eggs that provide all the resources for offspring development, why bother with males and with fertilization of those eggs? In particular, why throw away half your genes in egg production, in expectation of a male providing the missing half? Why not double your reproductive success by putting all your genes into your eggs and have them reproduce a new generation with your genotype? By various twists of evolutionary fate, often with hybridization with another species playing a role, some species have turned into exclusively parthenogenetic females. Their eggs always keep the maternal genotype intact and develop without any intervention by a sperm. Other species reproduce most often without fertilization, but males are produced and sexual reproduction happens whenever the next generation is to develop in an unpredictable environment: in another season after a long period of dormancy or, for parasites, in a new host.

WHY BE A HERMAPHRODITE?
·······

Once again I borrow from the same bountiful source; the question in this subhead is the title of a 1976 article by John Maynard Smith and two distinguished collaborators. It takes us to the next episode in the evolution of sexuality, and logically the next big question. Should each individual make both kinds of gamete, or should some make only sperm and others only eggs? If an organism makes both, we call it a hermaphrodite. If it makes just one, it is a male or a female. Living organisms give a great diversity of answers to the question of which kind of gamete should be made, and when. The *when* is needed because there are two kinds of hermaphrodite, the *simultaneous* and the *sequential*.

Simultaneous hermaphrodites make both eggs and sperm at the same time. There are many examples, earthworms and various snails being the most familiar animal examples. When they copulate, two individuals simultaneously fertilize each other's eggs. As a general rule, hermaphrodites do not practice self-fertilization, which would partly frustrate the original purpose, and perhaps some additional purposes, of sexuality. A few hermaphroditic plants may regularly self-fertilize, and many others can do so as a last resort. They fertilize some of their own eggs with their own pollen if they do not get an adequate supply from other individuals. Seeds resulting from self-fertilization are often less vigorous than those produced by outcrossing. They illustrate what is generally known as *inbreeding depression*.

Why are earthworms and many snails and most plants simultaneous hermaphrodites, while most vertebrates and insects are either male or female? As a general rule, an organism will be a simultaneous hermaphrodite if the adaptations for one sexual function can also serve the other. For example, a plant that makes flowers visited by insects is using those insects in two ways: to take its own pollen to other plants and to get pollen from them to itself. Nectar or large showy petals thus serve both its male and its female interests.

If male and female adaptations do not have this versatility, and especially if they would interfere or compete with each other, the sexes will be separate. Imagine an animal with strikingly different male and female adaptations, for instance, a deer. For every breeding season, a male must make a nutritionally expensive set of antlers and use them in sometimes lethal violence against other males in competing for sexual access to females. What good would this weaponry and bellicose behavior be for later female functions of pregnancy and lactation? From the female perspective, they would be an egregious waste of resources and unjustified hazard. Likewise, if a male needed to conserve energy for an impending pregnancy after the rutting season, it could seriously compromise his efforts to fertilize females.

None of the reproductive adaptations of deer look like mechanisms that could serve both male and female interests. Deer, and mammals in general, always have separate sexes.

But there are a few hermaphroditic vertebrates: some fish, like snails and earthworms and most plants, are simultaneous hermaphrodites. More often we find sequential hermaphrodites among the fishes. A young fish matures as a member of one sex, reproduces for a while, and then changes sex and reproduces as appropriate to that second sex. Some start reproducing as females and later change into males, and others do the reverse.

The direction of sex change always conforms to what is called the *size-advantage* model. Other things being equal, it helps to be bigger than your competitors. The bigger a female fish is, the more eggs she can make (and accommodate in her body); and the more eggs she makes, the more will be fertilized and convey her genes into the next generation. If she can produce 9,000 young by making 10,000 eggs, she can probably produce close to 18,000 young by producing 20,000 eggs. The same sort of size advantage works for males. The bigger they are, the more sperm they can turn out. Even if only one in a million ever fertilizes an egg, making two billion sperm will probably achieve more fertilizations than one billion could.

But the picture is not as simple for a male. The male that produces the most sperm may well fertilize the most eggs, but making twice as many is not likely to result in twice as many fertilizations. The reason is that a male is his own worst enemy in sperm competition. If a large number of sperm approach an egg and one succeeds in fertilizing it, many of the nearby unsuccessful sperm probably came from the same male. Females do not have this problem. If one egg gets a sperm, that is unlikely to deprive another egg, from the same female or even from another, of fertilization. There are usually more than enough sperm for nearly all available eggs to be fertilized. So the advantage of large size is likely to be greater for a female. Under these conditions, a sequential hermaphrodite should start reproductive life as a male and then, when it reaches the size of diminishing return for sperm production, change into a female and reap the benefits of unrestrained fecundity.

This argument assumes that males compete mainly by trying to outdo each other in sperm production. In some species there is a more direct and decisive contest competition, with a male taking exclusive possession of one or more females or forcefully appropriating a special location where females spawn. With this kind of mating system, only those males able to dominate other males and keep them away from fertile females have any success at fertilizing eggs. A sequential hermaphrodite in this system should start reproducing as a female and change into a male only when it has grown big enough to have a fighting chance in the mating contest.

The observations strongly support these ideas. Male-first hermaphrodites have what is called scramble competition for mates, often with several male-phase individuals releasing sperm while crowding around one, usually much larger, female-phase individual as she releases eggs. Female-first hermaphrodites always have contest competition, usually with large male-phase individuals occupying vigorously defended spawning territories that a succession of smaller female-phase individuals visit to lay eggs.

Why wouldn't these expectations also apply to mammals? A successful male gorilla is perhaps twice the weight of a female, and he keeps a harem of several females. Large size is important in enabling him to meet frequent challenges from sexually deprived competitors. Why do gorillas not mature as females at perhaps 80 kilograms, but then keep growing until they reach 160 kilograms and turn into males? The answer no doubt lies in the structure of the mammalian reproductive system. To replace a female mammal's uterus and vagina with a scrotum and penis and accessory structures would require some major rebuilding and probably take a long time, during which no reproduction could take place. By contrast, a female fish's reproductive system may consist of an ovary for making eggs, often less than 1 millimeter in diameter, and a simple tube for conveying them to the outside. Such an ovary may retain embryonic testicular tissue that can quickly grow and produce sperm that can exit by the same tube that once served the eggs. The sex-change operation is much easier and quicker for a fish than it would be for a mammal.

SEX RATIO
·······

I dealt with this question in chapter 3, as an example of a way in which natural selection clearly does not work for the benefit of the species. Here I will mention a few aspects of the problem in the context of the evolution of sexuality.

Consider the following experiment, with the most familiar of all organisms. You identify ten idyllic but uninhabited tropical islands, and you have five hundred young men and five hundred young women at your disposal. You stock one island with ten men and ninety women, another with twenty men and eighty women, and so on, to the last island on which you put ninety men and ten women. After ten years you revisit each island and count the children. Which island would you predict to have the largest number? Surely the one with the largest number of women. It is their capabilities that limit the number of babies that can be produced and nurtured for the first year or more of life. Fertilization by a man merely triggers a woman's reproductive process and makes no economic contribution to it. Ten randomly selected men will probably be quite adequate to keep ninety randomly selected women reproducing at what might be considered a normal rate.

So if reproductive efficiency were what determined the evolution of sex ratio, the number of girls born should greatly exceed the number of boys. The data, as everyone knows, are quite contrary to this expectation: the number of boy and girl babies are very nearly the same, with the boys actually slightly more numerous. It would appear that considerations of collective reproductive efficiency do not determine what sex ratio will be produced.

This conclusion was emphasized in chapter 3 as the classic example of frequency-dependent selection. The evolutionary expectation is that tendencies to produce offspring of the minority sex will be favored and accumulate, until that sex is no longer in the minority. When males and females are equally numerous, neither has a special advantage in mate competition, and sons and daughters are equally effective at providing grandchildren. At evolutionary equilibrium, if you have four children, it does not matter whether they are all boys or all girls or some mixture of the two.

With this sort of frequency-dependent selection, the adaptive value of a characteristic (for example, maleness or femaleness) depends on what proportion of the population has that characteristic. This kind of selection does not produce an adaptation to some environmental condition, but instead abolishes the condition, in this case some excess of either males or females. The result is that boys and girls are about equally numerous. The exact expectation, derived from careful consideration of sex differences in rate of development, mortality rates, costs to parents, and other factors is that selection will establish a population in which the parents collectively expend equal resources on sons and daughters. More broadly still, any population, whether of simultaneous or sequential hermaphrodites or with separate sexes, will be spending resources equally on male and female functions.

The expectations conform remarkably well to theoretical expectations, for both human and other well-studied animal and plant populations. This claim applies to life cycles quite different from the human, for instance, social insect societies in which most individuals are sterile females. It applies to hermaphrodites' sex-allocation ratios. Outcrossed simultaneous hermaphrodites will devote resources equally to male and female reproductive effort. Sequential hermaphrodites will time their sex changes in relation to how many male-phase and female-phase individuals are in the local population. The sex-allocation problem is one of the most satisfying applications of modern Darwinism.

SIZE OF MALES
·······

Even if we confine ourselves to species with separate sexes, and with males and females about equally numerous, the diversity of life histories is bewildering. Modern Darwinism is beginning to make sense of what formerly were considered just amazing curiosities: species in which males are sexually mature at birth, and those in which both sexes remain immature for decades; species in which females are much larger than males, and vice versa; fish and birds whose mothers always desert their young, which are tended with great solicitude by their fathers; species in which males and females pair only briefly, and others in which couples stay mated for many breeding seasons, perhaps for life.

Here I will discuss only one of many such problems, the relative sizes of the sexes and the factors that influence the evolution of this size dimorphism. I discussed one example, the relative sizes of male and female seals, in chapter 3. The expected and well-confirmed rule is that, where there is no sort of contest among males for mates, females will be larger than males; large size will be more important for females because of its direct effect on fecundity. There are far more species in which females are larger than the other way around, and the difference can be extreme. Imagine a species with a really sparse population in the open ocean. Individuals moving about at random seldom encounter each other. As is common in marine organisms, they start life as minute larvae, perhaps a fraction of a millimeter long, and grow to adults of several centimeters. One day a young, 1-millimeter male drifting in the plankton chances on a female ready to lay eggs. No other male, of any size, is around to compete. What advice would you offer that tiny young male?

My suggestion is: hang on to that female; you may never find another. And forget about growing up to be big and strong (and wasteful). Mature as quick as you can, before some competing male arrives and spoils what looks like an ideal situation. You need make only one microscopic sperm to fertilize each of those eggs from that female, and even a tiny male should be capable. If you start maturing now, and are ready with a few sperm cells tomorrow, you can fertilize a few of the available eggs tomorrow. As long as your production of microscopic sperm is adequate for those big, nutrient-loaded eggs, your reproductive success is assured. If another male does come along and competes for those eggs, you have at least made use of the splendid opportunity afforded by being the first to find that productive female.

This is the theory now invoked to explain the dwarf-male phenomenon found in many marine animals. The males of some species of barnacle are tiny rudiments tucked inside the shell cavities of the females. In some deep-sea anglerfish, the males bite into the skin of the larger females and hang on for the rest of their lives. In both examples the mature males are physically degenerate and live as parasites on the females, which may reach a thousand times the mass of their mates. Less extreme examples of the dwarf-male phenomenon, with males perhaps a tenth of the female size, are much more common and are found in all the major groups of animals.

Contest competition among males leads to a dramatically

different course of evolution, as explained in relation to sequential hermaphroditism. If you are a 9-gram male of a species in which males of 10 grams can almost always defeat those of your size, and if there are many 10-gram competitors around, it makes no sense to mature and try to win a mate. It makes sense only to keep growing. It should not take very long to grow from 9 to 11 grams, and imagine how much better a competitor you would be then. Under these circumstances, the evolution of ever larger body size in the male sex would not be surprising. This process apparently never produces the extreme difference in body size seen in the dwarf-male species. Some seals and primates and the sperm whale have males of several times the female mass, but never as much as ten times.

This chapter has dealt with many levels of the immensely important and complex topic of the evolution of sex: how the basic molecular processes of sexuality originated; the roles played by the resulting genetic recombination in life histories; the reasons why same-size gametes have been unstable and evolved into minute sperm and giant eggs; what determines whether and when an individual will produce eggs or sperm or both; and, for those species in which one individual produces only one kind of gamete, how it will compare with and relate to individuals that produce the other kind of gamete. The next chapter narrows this general focus down to the problems of sexuality in one species, our own.

CHAPTER 6

THE HUMAN EXPERIENCE OF
SEX AND REPRODUCTION

Organisms show an amazing diversity of life histories. They vary dramatically in such quantities as number of offspring per adult, time required to reach maturity, rates of growth and mortality at different stages. Our species is extreme in several such measures. Organisms also differ in the cellular details of the reproductive process, with two major modes, sexual and asexual. Here again there is great variety. Many organisms reproduce only sexually, some only asexually, and many more both ways, usually with a predominance of the asexual process.

We clearly belong to the exclusively sexual group, except for the occasional birth of identical twins, the result of an asexual division process by an early embryo. A twin birth happens so rarely, and until recently was so unlikely to produce even one successful offspring, that it can be thought of as a maladaptive abnormality. So we exemplify those organisms that reproduce only sexually, with separate males and females rather than hermaphrodites. We also show the effects of sexual selection in male-female contrasts that have no direct bearing on the production or nurture of young: the beards and greater size and mechanical strength of males, the difference in voice pitch, and other average contrasts.

PREGNANCY
·······

From the perspective of natural selection, what relationship could be more assuredly benign than that between mother and fetus? Producing and nurturing a fetus is a clearly adaptive way for the mother to get her genes into future generations. Her genetic success is crucially dependent on the successful completion of pregnancy. Similarly, the fetus is utterly dependent on the placental connection to its mother. Only through her can it get food and oxygen and rid itself of wastes. So isn't human pregnancy a condition in which we can confidently expect a mutually beneficial program of uncompromised cooperation?

No, it is not. Mother and fetus do have interests in common, but some facts I failed to mention should lead us to expect conflict. A woman, to reproduce successfully, has to nourish a fetus (ultimately, perhaps several such), but this need not mean nourishing the particular fetus she is carrying. Suppose it is seriously defective and likely to be born with a grave handicap in its efforts to survive and reproduce? If the defect is apparent to the mother (to some monitoring mechanism in the uterus), it may be adaptive for her to abort as soon as she can and wait for a more promising specimen. A substantial proportion of human embryos are aborted so early that the mother can have no awareness of ever being pregnant. Much of this could result from an adaptive selectivity of embryo retention by the maternal tissues. But mightn't it be adaptive for the embryo to frustrate this maternal selectivity? It is a life-and-death matter for an embryo to get safely implanted and to stay that way. It would be only a dubious quantitative gain for the mother to reject one embryo in the hope of soon having a better one. It is difficult to pick a winner in this sort of conflict. Winning is more important for the embryo, but its resources are limited compared to the mother's. The only certainty is the existence of conflict.

This expectation of conflict is often countered by arguments that have intuitive appeal, but are difficult to evaluate logically. The only way to achieve any real clarity is by formally considering prospects for the genes in the various individuals with interests at stake. There is not only a mother but a father, with genes just as heavily represented in the fetus as those of the mother. There are also other relatives with genes identical by descent to those in these three individuals. Especially relevant would be other children from the same mother, some perhaps already under foot, others just future possibilities, but very much relevant to the ultimate genetic success of the mother. We must also consider the interests of the father's genes in present or future children. All these possible children can be full or half siblings of the present fetus.

Imagine the ideal situation for the ultimate efficiency of the reproductive process from the perspective of one individual, the fetus. Its genetic programming for growth and organ development is best served by a certain input of nutrients. More is always better than less, up to some point of diminishing returns. As adults, we can appreciate that it is possible to eat too much, and this must be true for any stage in the human life history. The level of fetal well-being in relation to nutrient intake must be something like that shown by the curve in the figure. To be accurate for an evolutionary argument, the curve must show fetal inclusive fitness, not merely personal well-being. Its ideal is based partly on effects on close relatives, as explained for kin selection in chapter 3.

The intake best for the fetus is that which gives it the highest inclusive fitness, marked O_f on the diagram. Unfor-

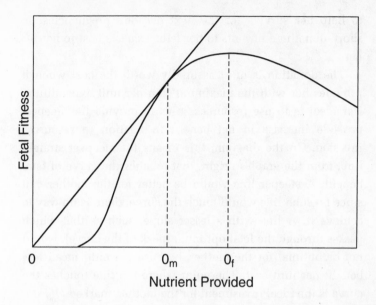

tunately for the continuance of this ideal, its achievement no doubt requires that many genes do things exactly right in development. Suppose at some gene locus a mutation happens that causes the mother to curtail slightly the concentration of some nutrient. It does not matter what the mechanism might be or through what developmental pathway it is achieved. Let's assume it causes a pregnant woman to curtail just slightly the concentration of glucose delivered to the placenta. Unless there is some compensating disadvantage, selection among pregnant females would favor this new mutation and lead to its partial or complete replacement of the ancestral allele.

To see why, we need to look at the diagram from the mother's perspective. Whatever she might donate to the current fetus is a resource that she will not be able to use for other purposes. Glucose delivered to the fetus cannot be used to provide energy useful in milling grain to feed to her four-year-old. It cannot be used to replenish her glycogen reserves for tomorrow's emergency or next year's lactation or to help her survive for the next decade's pregnancies. In short, donating nutrients to her fetus exacts a cost to her fitness.

The donation is most assuredly worth the cost when it provides her with the maximum gain per unit expenditure. Her ideal is to use resources so as to provide the steepest possible increase in net benefit in relation to resources invested. On the diagram, this means the steepest straight line, from the graph's origin, that touches the curve of fetal benefit. A steeper line would be better for the mother, but since no such line would touch the curve, there is no way to achieve it. A line with a lesser slope, such as that which passes through the fetal optimum (peak of the curve), would not be optimal for the mother, because it would mean less benefit per unit cost. The point where the line touches the curve is the ideal investment for the mother, marked O_m.

So, as the graph shows, the ideal for one participant in this interaction and that for the other are different. Selection acting on fetuses will favor greater fetal nutrition than that acting on mothers. The mathematical details of this sort of selection are complicated, but the expected result is a compromise, a population in which maternal delivery of a nutrient, such as glucose, averages a bit less than the fetal ideal and a bit more than the maternal. A compromise surely, but not necessarily a placid one. As long as the prevalent value is different from the ideal for any individual, selection on such individuals will tend to move the value in the direction of the ideal. Mothers will be selected to decrease and fetuses to increase the glucose supply. Any progress for the maternal stage will be met by countermeasures by the fetal, and vice versa.

An erroneous objection to this reasoning is sometimes heard: we cannot talk of selection on mothers and selection on fetuses as if these were different sorts of individuals in a population. They are just different stages in the life cycle. Whatever fetuses might gain from a nutrient level that exceeds the maternal optimum would be exactly compensated by losses to mothers carrying such fetuses. This argument would be valid if reproduction were asexual, but it ignores two important aspects of human sexual reproduction. First, only female fetuses can ever live to experience pregnancy. If a gene causes all fetuses to extract extra nutrients from the mother, all will benefit, but only half can later pay the cost of this benefit. The second factor is simply that sexual reproduction makes human inheritance Mendelian. If the newly mutated allele A in genotype Aa augments the fetal greed, only half the later offspring of that fetus will bear the A. So only half the daughters (a quarter of the offspring) of a greedy Aa fetus will inherit the A and pay the cost. Those getting a will be less demanding. The reasoning here was worked out in 1974 by the American biologist Robert L. Trivers for a more conspicuous sort of conflict in mammals generally, that seen (and often heard) between mother and young at weaning.

There really is evidence of the same kind of conflict between mother and offspring before birth. In a landmark contribution, the Harvard biologist David Haig examined the phenomena of pregnancy from an evolutionary perspective and found pervasive evidence of intense conflict. The first thing a fetus does on implantation is to produce a placenta that invades the uterine tissues and disables the local control mechanisms of blood flow. Thereafter the mother will not be able to control the blood flowing to the placenta by constricting the arteries that lead to it. A bit later the placenta, which is formed by the fetus for its own nourishment, will start secreting chemical stimuli into the maternal blood. One of the chemicals stimulates blood vessels elsewhere in the mother's body to contract and raise her blood pressure. This indirectly results in still more blood being diverted to the placenta. Other chemicals have an action opposite to that of insulin and increase the sugar content of the blood.

The mother, as expected, does not simply acquiesce in this manipulation. She makes chemicals that act against those produced by the fetus. The fetuses are thereby selected to evolve still greater production of their chemical manipulators, and so on in a grossly escalated chemical arms race. An extreme example is provided by the hormone *human placental lactogen*, which opposes the action of insulin and is always present in tiny traces in human tissues. It may be produced by a fetus at a rate that raises its concentration, in its mother's blood, to a thousand times that normally found in a nonpregnant woman. As Haig points out, raised hormone levels, like raised voices, are a sign of conflict.

These are not matters of merely intellectual interest but underlie some serious obstetrical problems, and one of Haig's clever analogies makes it clear why this should be. Each participant in the conflict is trying for only a slightly greater or slightly lesser setting of some physiological quantity, such as blood pressure. This is analogous to your wanting the radio slightly louder and your companion wanting it slightly quieter. You start pushing the setting up and your fellow listener responds by pushing it down. Such persistently frustrated effort makes each of you push ever more forcefully, as long as the current volume is even slightly different from what you want. But suppose one of you quits or is suddenly incapable of pushing. The other's forceful effort will now make the sound either so loud as to be painful to both individuals or disappear entirely.

This is analogous to the occasional pregnancy in which a mother is seriously deficient in her ability to resist some form of fetal manipulation. A possible consequence is a harmful hypertension, in extreme cases the dangerous condition called eclampsia. Another is gestational diabetes. In a sense these diseases are indications that the fetus has won the contest, but such winning can be as bad for the winner as for the loser. Causing serious damage to its mother cannot be in a baby's interest. Even a mild preeclampsia may cause blood clots that occlude placental blood vessels and reduce rather than augment fetal nutrition. A complete victory resulting in maternal death would be a total loss for the fetus.

Genetic conflict provides a rationale for the recently discovered phenomenon of genetic imprinting. Ordinarily the effect of a gene is in no way dependent on where it came from. When Gregor Mendel, the nineteenth-century founder of the science of genetics, pollinated red-flowered peas with pollen from white-flowered peas, he got the same results as when the pollen producers and recipients were reversed. This need not be true in mammalian reproduction, as has been known for a long time: mules and hinnies are not exactly the same. Recently it has been found that mouse genes may be transmitted to eggs and sperm in a turned-on or turned-off state. Males make sperm with turned-on genes that extract greater nutrient delivery from the maternal blood. These same genes transmitted by the female are turned off in the egg, and the egg turns on genes that act in opposition to the turned-on sperm genes, and so on.

The volume-control conflict is again analogous. If any male did not turn on the genes that oppose the maternal interests, the maternal turned-on genes would overwhelm the fetal attempt to nourish itself. The mother would completely win this conflict with her mate, with the result that both his and her own reproduction would be frustrated. This phenomenon of genetic imprinting has not yet been shown conclusively in human heredity. We cannot look for it in experiments on human subjects with the facility that we can with mice. Yet there is no reason to doubt the generality of this effect in mammalian reproduction or its occasional role in obstetrical malfunction.

Fortunately for her peace of mind, a pregnant woman is unaware of the chemical warfare she is waging with her fetus. If she wins the conflict, she aborts the fetus and is unlikely to know why. If she loses, she will be among that small minority of women who suffer seriously from gestational diabetes or other obstetrical ills. If the usual stalemate is achieved between the warring parties, all will go well with her pregnancy and childbirth.

Unfortunately for her peace of mind, a newly pregnant woman will probably be acutely aware of symptoms normally associated with poisoning or stomach pathology. She will have persistent nausea, sometimes acute enough to lead to vomiting and reduced intake of many foods. What could be going on here? Why should symptoms of illness so regularly accompany an essential part of human life history? Curiously enough, this obvious question was entirely ignored by biologists and medical specialists until 1988, when Margie Profet, then at the University of California, made her radical new suggestion: the nausea and other symptoms result from a recalibration of the brain mechanism that detects and reacts to toxins circulating in the bloodstream. Early in pregnancy it reacts to hormonal changes and resets the nausea threshold. It becomes sensitive to minute traces of toxins, most often those evolved by plants to protect themselves from herbivorous animals. These minute traces are readily tolerated by maternal but not by fetal tissues. In the rapid organ formation and tissue specialization of the first three months of pregnancy, the fetus is readily harmed by toxin concentrations that would not cause problems for adults. Morning sickness seems to be an adaptation to protect the fetus from toxins that might interfere with normal development.

In 1992 Profet greatly expanded the theoretical discussion and mass of evidence in favor of her interpretation of morning sickness, and in 1995 made it a key idea for her book *Protecting Your Baby-to-Be*. If her idea is correct, it has serious medical implications. Use of drugs to suppress the nausea and related symptoms, such as specific food aversions, could increase the likelihood of birth defects.

CHILDBIRTH
·······

The timing of childbirth is another phenomenon in which we might well expect conflict. I presume that the womb is a safer place for a full-term fetus than is the great outdoors, and the longer it stays there the better nourished it will be when it first has to cope with the external world. The mother may find an earlier delivery more convenient. It would give earlier relief from her mechanical burden, and a younger and smaller baby would mean an easier labor and delivery. So she might be better served by a slightly shorter-than-average pregnancy, while the fetus might be better off if it were slightly longer than average. The conflicting ideals would be only slightly different, but this is true of conflicts during pregancy. As she gets closer to delivery, the mother produces oxytocin, a hormone that initiates childbirth. The difference in genetic interest between mother and fetus provides a reason for predicting the existence of a fetal secretion that counteracts oxytocin.

The current evolutionary insight that most clearly relates to human childbirth has nothing to do with conflict of interests between mother and fetus. It relates to an unfortunate human legacy from the remote past. When the pelvic ring of bone was first evolved by early land-dwelling vertebrates, all systems exiting the body—digestive, reproductive, and excretory—passed through. The same basic geometry is retained by all modern descendants. Take a close look at the next human skeleton you encounter. Note the bony circle of left and right pubis in front and left and right ischium and their vertebral connection behind. Babies have to squeeze through a narrower space than that, because the ring is crowded by the vaginal wall and rectum and other structures. Birth is more difficult for a woman than for most other mammals—it's such a tight squeeze.

Now look at the space between the pelvis and the ribs

and sternum: a great wide, boneless gap. Why not give birth through that ample opening? In fact many women do, those in cesarean deliveries. A surgeon makes the opening that our evolution failed to provide. It is the preferred opening in this one important respect: it can be of the most convenient size, so that no mechanical problems need arise in getting the baby through. In all other respects, of course, it is less desirable than the vaginal passage through that tight pelvis. This general vaginal superiority does not alter the fact of its serious design flaw, its conveyance of the newborn through a stressfully narrow constriction. Any sane engineer would have provided the vagina with a natural opening in the lower abdomen that would be superior both to the one provided by evolution and to that provided by a surgeon. (There are many other basic design flaws in the human body, some of which are discussed in the next two chapters.)

CHILDHOOD
······

Surely I need offer no evidence for the proposition that children can be selfish in their dealings with parents and with one another. The human weaning conflict can be intense, and is especially evident for that minority of modern women who breast-feed their babies to an age of advanced language capability. A three-year-old can be not only loud but impressively eloquent in offering reasons why Mommy should continue nursing. This is merely an early example of many intense verbal arguments to come.

Much of the conflict within families can be interpreted from an evolutionary perspective as based on genetic selfishness. The graph earlier in this chapter applies equally to glucose delivered to the placenta as to a weekly allowance of spending money. The desire for a more lavish allowance is merely a special example of the more general and biologically understandable desire for more of whatever seem to be valuable resources. Throughout most of human evolution, more resources led to better health and the higher social status that would have led to more desirable or more numerous mates and allies and providers.

Children in the years between birth and puberty are not, in a normal sense, engaged in sex and reproduction. Yet success in childhood—success in getting adequate food, in avoiding illness, in forming valuable friendships with other children and adults, in learning what it takes to succeed throughout life—is of vital interest to ultimate biological success in transmitting one's genes. Sometimes the pursuit of different aspects of fitness will be in conflict, and it may be difficult to work out the ideal compromises and trade-offs. Perfecting one's skill at making a stone tool can decrease the time available for digging clams.

Such problems can be especially difficult in today's grossly abnormal environments. Learning how to solve quadratic equations may take time from learning how to play a nocturne or pass a soccer ball or weed the strawberries. Success at acquiring resources and social status and helping your relatives is no longer a secure path to ultimate genetic success. Such modern worldly attainments may in fact be associated with reduced genetic success, and socioeconomic failures may raise more children. Yet the motivations that lead to skill at soccer or mathematics or producing admirable gardens are no doubt based on their historical usefulness in passing genes into future generations.

MATING AND PARENTING
·······

Human growth and maturation are slow, but there comes a time when those who were children must be admitted to adult society and its privileges and responsibilities. The process may be gradual and informal, but often there are social rituals that mark a formal passage to adulthood: puberty rites, graduation from high school, marriage ceremonies, and the like. Always the young adults enter a complexly competitive mating game. This is a crucial kind of competition for resources, because mate quality and family connections have through human history been important for the achievement of genetic success.

Human sexual competition has recently gotten the attention of biologists intent on understanding the evolutionary origin of current human motivations. Because of the clear contrast between human male and female reproductive physiology, male and female sexual strategies, as in mammals generally, are expected to differ in important and somewhat conflicting ways. A woman must do all the nurturing of her offspring during pregnancy, and most of it thereafter during lactation. This physiological burden strictly limits the number of her weaned children and does not depend on which man or men supplied the paternal half of the offspring genotypes. Her productivity is set by her own physiological capabilities, and these in turn depend on her personal well-being.

Successful weaning of a child is only part of the job of getting genes into future generations. The child has to be reared and equipped for competition for resources, such as mates, in adult life. It is mainly (but not entirely) for this postweaning nurture and protection and training that it matters what man or men may be available to help her to achieve genetic success.

Imagine yourself a newly nubile woman named Sarai, one of a few dozen nomadic hunter-gatherers in a closely woven social structure. You know many of your fellows to be related to you as siblings or cousins or aunts or nephews. Very few do not have some known genealogical or marital connection. There are other such groups within a few days' march, all rather similar in language and custom, and mem-

bers of these groups are sometimes available as marital part-
ners for members of your group. There is occasional violent
conflict with these others, but more often there is merely
benign competition and contacts for trade and brief social
interaction. Beyond these familiar and understandable people are others, with strange speech and many senseless or
repulsive customs. Fear and enmity are adaptive feelings
toward them.

Stone Age marital customs no doubt varied widely, as
they do today, but I will assume that you (Sarai) have some
influence on who becomes your husband. No doubt your
parents or other responsible adults are also part of the decision making, but their choice may be influenced by yours.
Whom do you choose? Ideally your husband should be
healthy, wealthy, and wise. He should be young enough to
have some likelihood of being around to help for another
decade or two. He should have the strength and skills and
influence needed to protect you and your children from
harm. He should have friends and relatives who will help
him in times of need. He should have a benign and stable
personality and be unlikely to abuse you or desert you for
someone else. Natural selection will favor those women
whose intuitive responses to men provide reliable evalua-
tions on these and related attractions.

The choice is exceedingly difficult because it requires, in
effect, predicting a most uncertain future: a man's future
health, longevity, social status, sexual inclinations, treatment of children not yet born. Even the best possible choice
may be a poor one because there are so few contenders. In
that band of a few dozen, how many eligible bachelors are
there likely to be? And need he be a bachelor? Seth, the
chief's son, a few years older than you, already has a wife
(perhaps pregnant or lactating) and is in the market for
another. Would marrying him, and becoming a second and
rather subordinate wife, be better than marrying that fellow
Abram? His socioeconomic status is nowhere near that of the

chief's son, but you would not be sharing him with another woman. And then there is that handsome, clever guy Kenan in the Eagle Clan whom you met last fall at the pomegranate festival. Might it not be better to wait and ask Uncle Enoch to open negotiations with Kenan's parents at the next festival? But waiting might prove disastrous. Cousin Jared's daughter is growing fast and looks like a possibly formidable future competitor for the favor of any of those possible husbands. You really do have a difficult problem to resolve, and so much of your future happiness depends on your making the right choice.

Now suppose you are that young man named Abram who needs a wife. Your genetic interests would be best served by a woman who is unburdened by any prior children that might compete with yours. This almost certainly means someone quite young. Youth also is predictive of a long future career of fertility and competence in rearing your children. The same would be indicated by any evidence of health and any personality features associated with willingness to assume the burdens of wife and mother. In evaluating prospective wives, these essentially personal features are more important than the socioeconomic status that was most crucial in Sarai's decision making, although obviously family connections are still important.

There is also an immensely important aspect of your genetic interests that was of little concern to Sarai. Your reproduction is not limited by your physiology. If you are normally fertile, the number of your children will be mainly a function of how many women you can inseminate. For you, the variation in quality among prospective wives may well be less important than their number. It is not likely that you will get more than one wife now, but you need to keep your options open and be alert for opportunities for additional wives in the future.

Let's say Sarai and Abram get married. The expected differences in reproductive strategy will extend without change into their years together. If both are normally fertile, Sarai will soon be pregnant. If the pregnancy fails, she will soon be pregnant again. If she produces a normal baby that survives perinatal and later hazards, she will nurse it for at least two and perhaps four years. Lactation will inhibit ovulation, and there will be no new pregnancies during this period. When her baby is weaned, she will soon ovulate and be pregnant again within a few cycles. If all goes well, this pattern will continue. At menopause, if she is among the lucky few who survive that long, Sarai will have produced four or five children, but probably no more. If Abram remains fertile and is considerate and proves a worthy husband and father, Sarai will continue to count herself fortunate and avoid any behavior that might arouse his jealousy. It is especially important for her to avoid giving him any reason to doubt that her children are his.

Abram's attitude will be rather different. Sarai's several babies, if most of them survive, can make him a winner in the reproductive competition in which he is intensely if unknowingly engaged, but there are many degrees of winning. If he had another wife of similar fertility, his winnings could be twice as great. If the opportunity presents itself, we can expect him to add another wife to his family. No such benefit from having an additional husband would accrue to Sarai, whose fertility is set by her own physiological limitations, not by the number of men who supply sperm. Another option for Abram may arise in an opportunity to inseminate someone else's wife. An illegitimate child will carry as many of his genes as any child of Sarai's, but if another man bears the economic burden, the extra child will not be a drain on Abram's resources.

It takes two to commit adultery. In any given society we would expect to find more men actively seeking extramarital matings, but women like Sarai do in fact collaborate. What is in it for her? Two possibilities. Suppose she has been cycling for a year or more with no sign of pregnancy, and she thinks she is thereby slipping behind in the tribal esteem. Adultery with a demonstrably fertile man might well end her current unproductiveness. She would no doubt want maximum assurance of secrecy, if Abram's suspected infertility is his only important fault. More likely Sarai's adulterous inclinations would stem from a more socioeconomic dissatisfaction with Abram. She might even openly assert that she would rather join Seth's harem than go on living with that loser she has been stuck with.

Women can be jealous too, and Sarai would no doubt be angry and unhappy at learning that Abram had occasional trysts with Jared's daughter. Yet his adultery is less of a threat to her than hers would be to him. The act itself is no threat, because Abram's ejaculate is easily replaced, and even her rival's pregnancy would be of little direct concern to Sarai. It would be a problem only if Abram were inclined to devote resources to the illegitimate child and thereby deprive Sarai's. Abram's adultery could also be an indication that he is not really satisfied with present arrangements. By contrast, Sarai's adultery could imply analogous problems for Abram and be a direct biological threat besides. If she is pregnant by another man, she deprives Abram of years of her contributions to his genes. If he is deceived and thinks the child his, there will be the additional loss of whatever resources he provides for that other man's genes in place of his own. Cuckoldry is a major threat to a man's genetic success, and it is expected that adaptations will be evolved to minimize its likelihood. So male jealousy is expected to be intense, hardly a new idea. It has always been intuitively appreciated, as shown by the many fictional portrayals of violent male jealousy. Remember what happened when Helen ran off with a traveling Trojan, or when Othello thought he had evidence of Desdemona's infidelity? The fictional accounts have all too many parallels in contemporary life. When women are murdered, it is usually by jealous men.

These ideas conform to most people's experience of and

intuitions about the relations between the sexes, but this is no basis for scientific acceptability. Fortunately, in the last two decades, various biologists, psychologists, and anthropologists have been gathering more respectable data. There is now abundant evidence on sexual conflict and sex differences in mate choice and marital morality. The biological basis of who murders whom in modern communities is detailed in the 1989 book *Homicide* by the Canadian psychologists Martin Daly and Margo Wilson. The classic works on the biology of human sexual attitudes are by two California anthropologists, *The Evolution of Human Sexuality* (1979), by Donald Symons, and *The Woman Who Never Evolved* (1981), by Sarah B. Hrdy. More recent works are cited in the endnotes.

....

CHAPTER 7

OLD AGE AND OTHER CURSES

Death is a prominent fact of life. If some accident or disease or violence does not kill you first, something called old age apparently will. There seem to be obvious analogies in the implements we use. A washing machine that has given years of faithful service suddenly fails. You summon the repairman, and he says the machine needs a new regulator relay. It costs $20 and his labor $80. You pay him $100, he replaces the troublesome part, and the washing machine once again washes clothes. Or perhaps, for a machine that old, you think $100 too much. If the cumulative wear and tear on the regulator relay caused it to fail, it is likely that other essential parts will soon fail. So perhaps you prefer to buy a new washing machine.

Isn't the human body much the same? If an essential part gives notice of impending failure in the form of alarming symptoms, you seek help from a repairman, called a doctor in this context. She fixes the faulty part, or replaces it (artificial heart valve, plastic hip joint). *Buy a new body* is not an option, but otherwise the washing-machine analogy seems valid. Sooner or later, from chemical deterioration, mechanical strains, or frictional wear, a part will malfunction. Since all parts are subject to some sort of wear and tear, other failures are not likely to lag far behind the first. Sooner or later the wear and tear will win. A doctor can no more keep you alive indefinitely than a repairman can keep a washing machine washing clothes indefinitely. This sounds logical, doesn't it?

No, it does not. The human body's decline of adaptive performance and increasing likelihood of death are nothing like the wearing out of manufactured machinery. Both the human body and a washing machine are subject to cumulative effects of age, but in quite different senses. The *aging* of an organism is a misleading synonym for *senescence*, which is by definition a deterioration. Aging can mean steady improvement for a bottle of wine or a slab of cheddar. In thinking of senescence, the analogy to the wear and tear, or corrosion, or other process that ultimately causes an artificial device to fail is utterly misleading. A washing machine is an object, probably made mostly of an iron alloy. Almost all the iron that was there at the beginning is still there ten years later, in exactly the same places it was originally. The machine, once assembled, has almost exactly the same size and shape it will have ten years later. Changes, such as wear and corrosion, accumulate with usage or merely the passage of time. None of this is true of the human body.

An organism is not so much an object as a place where various processes operate. Every time you exhale, you expel about a milligram of carbon atoms that, shortly before, were functioning parts of your body. How much do you lose every day by respiration, excretion, bleeding, and so on? About the same amount you take in. If your weight is a steady 50 kilograms and you consume 1 kilogram of food and drink every day, you must get rid of 1 kilogram of matter, or you will not maintain that steady weight.

You may imagine that automobiles eat gasoline and breathe air and exhale steam and carbon dioxide and traces of less innocent substances, but this is quite different from your intake and output. The gasoline and oxygen never become part of the car. There is never any doubt about where the fuel tank stops and the gasoline in it begins. Likewise, the oxygen sucked through the air filter goes through the carburetor into the cylinder, combines with gasoline, and goes out through the exhaust system. By contrast, some of the water and alcohol and sugars in a sip of wine, within seconds, have become parts of the machinery of your body. Starches and proteins and other, less soluble substances take longer to be assimilated and used, but that is what happens to most of them. Within an hour of your finishing a meal, components of that meal are vital parts of your nerves and blood corpuscles and muscles. Other such parts, in that same hour, have been removed from such roles and exhaled or relegated to your bladder or colon to be cast aside. No man-made machinery works this way. The machinery and its resources are always clearly distinguishable, and the machinery is not constantly remaking itself, as the human body is.

Of course, human parts are subject to wear and tear and chemical attack. Every time you touch something, some skin cells wear off. The collar on a white shirt that has been tight around your neck all day will be discolored, mostly by dead cells that have rubbed off your neck during those hours of friction. Fingernails are worn away by friction, as is shown by using a nail file. Teeth wear down, and are damaged by other mechanical stresses that can chip them away. Yet none of this has much to do with senescence. The skin and nails and other such tissues, such as the linings of digestive, respiratory, and other systems, are replaced as they wear away. The tires on your car wear away also, but that wear is not replaced by any new rubber formed within the tires. With every second of travel, there is less rubber there.

An old person's skin differs from a youth's not because it has been worn away but because it is less effective at replacing what wears away and in maintaining temperature and healing wounds and in all the other ways it serves to main-

tain well-being. The material in old skin is still quite temporary and will be mostly gone in a few days or weeks at most.
The same is true of nerves and blood and muscle and all
other tissue.

A candle flame is a better analogy to an organism than is
any artificial machinery. The heat of the flame first melts
and then vaporizes the wax. The vapors ignite and become
so hot as to give off visible radiation. The hot vapors in
much cooler air rise rapidly as they complete their combus-
tion and cool down. Below a temperature of about 250°C
they are invisible and no longer qualify as part of the flame.
The flame has a complex structure, with several recognizable regions, and with different components of the fuel burn-
ing with different rates and temperatures and giving off different spectra of radiations. The component materials are
there for a few milliseconds and then are gone and replaced
by other materials from the wax and the air being consumed.

This analogy has its limitations because an organism is
far more complicated than a flame, and the rate at which different substances are taken in and given off varies enormously. A carbon or hydrogen atom in a metabolically active
molecule in one of your muscle or gland cells will probably
be somewhere else within minutes or hours, and out of you
entirely within a few weeks. If you are fifty years old, a calcium atom in one of your teeth may have been there forty
years ago, and it may still be there when you are ninety, if
you are still alive and still have any real teeth. But teeth are
exceptional. Most tissues, even bone, have appreciable rates
of material replacement over a period of weeks or months.

So you are not like a washing machine or a car and are
not defined by the material present at this moment. You are
a complex system of activities that makes temporary use of
various kinds of matter, but that matter is not you. You, and
all other organisms, are continuous systems of material flux,
of matter moving in, playing a role, and moving out. You are
more like a candle flame or a whirlpool than like a washing

machine. Our bodies are subject to wear and tear, but these processes are normally countered by evolved adaptations, like the steady replacement of cells in skin and other tissues.

What about our adult teeth? No, they have limited regenerative capabilities, and their enamel cannot be replaced. This was a serious problem with the abrasive foods eaten by some of our ancestors. By the time they reached what we call middle age, their molars may have been worn down to gum level and their chewing seriously impaired. Yet this frictional wear is just like any other, of the fingernails, for instance. The important observation is that adult teeth, unlike fingernails, cannot be naturally replaced.

Why not? Elephants can live a long time on highly abrasive diets. When their molars wear down, they drop them and grow new ones. They can have as many as six sets of molar teeth during their adult lives, although nearly all die before using up six sets. Our developmental programming assumes that, on an average Stone Age diet, we will not live long enough to need more than one set of adult teeth. Does this mean that we are programmed to die before our teeth wear out? No, our deaths are not programmed; they merely become ever more probable as we get older. Dying is not something we do, it is something that happens to us at an age-dependent rate. That rate today is lower at every stage of life than it was at any previous time in human history, and far lower than in the historically normal conditions of the Stone Age.

WHAT IS SENESCENCE?
······

Senescence is a continuous decrease in the precision of control over the material flux on which we depend. An analogy might be made between your control of the balls in a billiard game and of the molecules in your body. With the cue, you might impel the cue ball on a course that will make it strike the eight ball and make it, and the rebounding cue ball, behave in ways conducive to your winning the game. With that collision, your precision of control decreases markedly, and my analogy becomes less apt. The billiard balls would be more like the molecular machinery of your body if they were equipped with sensors to inform them of any deviation from the ideal paths, and with mechanisms for correcting any departures from the ideal. The analogy would be still closer if these cybernetic feedback loops were themselves subject to deterioration as the game went on.

The start of your life was not like the cue striking the cue ball in a billiard game. The cybernetic controls of the machinery of your body were set in motion at the beginning of your embryonic development and will continue until death. When anything starts to go wrong, some sensory device perceives the trouble and sends a message to machinery that acts to set things right. If sunshine strikes your skin cells, with possible overheating or damage from the ultraviolet component, heat receptors may notify your brain, which then stimulates muscles to act in ways that cause you to walk to a shady tree. The skin cells themselves can perceive the problem and respond by making the pigment melanin, another way of putting vulnerable cells in the shade. If blood sugar drops to a functionally deficient level, it is perceived and corrective measures are taken by the conversion of glycogen or fat reserves into sugar, or perhaps just by eating. The list of our cybernetic control loops could go on and on, but as we age, our internal control mechanisms become ever less precise, and permit a cumulative breakdown of our evolved adaptations. The victim of this process is ever more sadly aware of declining performance, manifested, for example, in not being able to run as fast or see as clearly as was possible some years before. Sooner or later an essential system fails. Should we think of this as death from old age?

No, we should not. Old age is never itself a cause of

death. Everyone dies from some lethal challenge to the evolved adaptations. Senescence plays its part by making death from many kinds of challenge ever more likely, but there is no recognizable death process. A man at age eighty-five may die because his heart could not deliver the minimal supply of oxygen needed by his brain, while at age seventy-five this level of cardiac performance was easy. A woman of forty-five may die of obstetrical complications that she would have avoided easily at thirty-five. A man or woman at thirty-five might be killed by a lion from taking a half-second too long to climb a tree, when the needed speed would have been achievable ten years before. This is how senescence is related to death. Death from many causes becomes more likely as we get older, because our life-preserving adaptations irreversibly deteriorate. There is no death mechanism, no objectively definable life span or longevity. Everyone will live until killed by something.

Surely this picture of human development and deterioration cries out for an explanation. From a microscopic nothingness, we achieve a near maximum of physical and mental and reproductive capability in about twenty years. This mature but youthful human body is an endlessly complex and precise set of adaptations that biologists are really just beginning to appreciate. But now, after the seemingly miraculous production of that superb machinery, the body proves incapable of merely maintaining what is there. The youthful competence and vigor gradually yield to ever greater debilitation with ever higher probability of death.

AN EVOLUTIONARY THEORY OF SENESCENCE
······

The currently accepted theory builds on two biological concepts, historically normal *survivorship* and historically normal *reproductive value*. Survivorship is the average proportion, over a long period of time, of newborns who live to a given age. For those hundreds of thousands of years that made the final adjustments in human nature, as we now find it, it is generally conceded that perhaps half the infants reached sexual maturity, at about fifteen years. This and adult mortality were no doubt widely variable, but for the next twenty years or so, survival may have averaged about 96 percent per year.

At that rate, about 23 percent of the babies would still be alive at age thirty-five and 10 percent at age fifty-five. Senescence would have an ever more noticeable effect after age thirty-five, so that probably far fewer than 10 percent of the newborns would reach fifty-five. Survival to age sixty in the Stone Age would be an unrealistic expectation. If this kind of schedule of survivorship prevails for long periods of time, it should be obvious that fitness at age sixty or later would not be important to natural selection.

The other essential evolutionary factor, reproductive value, is defined as expected future reproduction for someone of specified age. The reproductive value of a newborn was rather low in the Stone Age because of the high likelihood of death before reaching maturity. With the approach of puberty, reproductive expectations went up because an individual was ever more likely to reach the beginning of a reproductive career. At puberty the reproductive value is close to its peak because the individual has in fact reached that crucial age, with the whole reproductive career, including the years of maximum fertility, lying ahead. It is the individuals with at least moderately high reproductive values that will be doing most of the reproduction in the near future. Obviously these ages are more important to natural selection than those of low reproductive value, such as what we call middle age. In theory, natural selection is equally influenced by survivorship and reproductive value, so that its effectiveness in maintaining adaptation is measured simply by the product of these two factors.

Suppose you could go back to the Stone Age and endow some human population with eternal youth without otherwise altering its conditions of life. Young adults have the realistic survivorship of 96 percent per year, and this rate continues forever, because of their eternal youth. Those few who are alive at age 100 will still be as youthful and fit as they were as young adults. The upper figure on page 121 shows the survivorship curve for this population, and also reproductive value for each age. Since eternal youth assures continued high fertility, reproductive value, once it reaches its peak value, stays there ever after. At age 100 the expectation for future reproduction is the same as it was at fifteen. This maximum value has to be *four* to maintain the population, because only one of every two will survive to the start of reproduction (here set at age fifteen), and because it takes two (male and female) to produce each infant.

The lower figure shows the crucial product of survivorship and reproductive value in this senescence-free population. It is correct only with the mathematically convenient simplifications that the population maintains a constant size and that full fertility is reached immediately at puberty. Prior to that, increasing reproductive value exactly compensates for decreasing survivorship, so that the product stays the same. Thereafter the reproductive value remains constant, because there is no senescence, but the survivorship and the product of the two terms decline exponentially.

The important lesson from this fanciful Stone Age population is that it would quickly lose its eternal youth. Natural selection is always busy weeding out genes that decrease their carriers' lifetime genetic success and spreading those that increase this crucial score. Suppose a mutation occurred that caused a 2 or 3 percent increase in fertility or other improvement for individuals in their teens and twenties but made them become extremely vulnerable to liver cancer at age 100. Half the individuals would enjoy at least some of the benefit, and a quarter would get it all, but fewer than 2

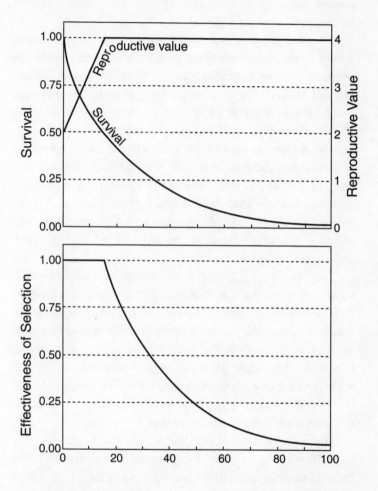

Top. Age distributions of survivorship and reproductive value in a senescence-free population with survivorship of 96 percent per year for adults, and sexual maturity at fifteen years. A fifteen-year-old in this hypothetical population with eternal youth is less likely than we are to reach what we would consider a normal life span, such as seventy years. It would be more likely to reach one hundred or older.

Bottom. Effectiveness of natural selection (product of survivorship and reproductive value) as a function of age in this same population.

percent could pay the fatal cost. The average benefit would exceed the average cost, even though centenarians would have severely reduced fitness.

All genetic variation, either existing or newly arisen, would be evaluated according to the obvious criterion: What proportion of the individuals would be affected by each benefit and cost? The process would be constantly acting to bias adaptation in favor of early ages at cost to the later ones. There need not be any actual early benefits for senescence to evolve. As long as selection is more effective at suppressing unfavorable effects early in life than later on, adaptations early in life will be more effectively maintained than those at later ages. So that eternal-youth population, if it ever existed, would be unstable. Natural selection would quickly enhance youthful fitness at the expense of old age, and senescence would soon evolve.

The abstract idea that survivorship and reproductive value determine the effectiveness of the evolutionary process in maintaining adaptation does not imply any simple observational rules, such as how a thirty-year-old would compare with a forty-year-old in fleetness, fertility, or vulnerability to infection. Any such measures would depend on special aspects of the environment in which they are measured, and the values for any current population would surely be quite different from those in the Stone Age.

There is also another complicating factor. In maintaining high fitness at age twenty, for instance, selection may favor features that incidentally increase fitness at later ages. For a technological analogy, suppose you manufacture a product to be sold with a one-year warranty. The quality of this product after eleven months is obviously more important to you than that after thirteen months. Yet it is to be expected that trying to make something that will, for sure, last twelve months will often result in its lasting a lot longer than that. Your products will not all fail on the 366th day. Similarly, selection in the Stone Age for maintaining fitness at age thirty, or at least slowing the loss of it, might be expected to result incidentally in sixty-year-olds able to live healthily in our benign modern environment. Still, we have learned not to expect too much of sixty-year-olds. They are not as fleet, not as fertile, and not as good at avoiding Hodgkin's sarcoma and other forms of cancer. We would expect them to have higher mortality rates, and they do. Modern death rates at age sixty are utopian by historically normal standards, but they are about ten times what they are at thirty and continue to increase ever more steeply. Fully a third of the people reaching 100 will die before reaching 101.

Senescence, a steady decline in adaptive performance starting with the attainment of full maturity, is thus an inevitable aspect of human nature. It makes death ever more likely, but this does not mean that there is a normal age of death. There is no programmed *natural death,* nor any characteristic life span for the species. The age distribution of mortality rates results from an interaction between the evolved rates of senescence and the rigors of the environment in which the living takes place. The maximum life span of our own or any other species is determined by this interaction and the size of the sample observed. The maximum age in a sample of a million is likely to be much greater than that in a sample of a thousand. A sample of a billion will give a still higher maximum longevity.

The study of senescence has been dominated too long by actuaries and others preoccupied with death. Its proper study is in measurable aspects of the vital functions of individual adults through a succession of years. Deaths are not the data to be collected but merely the events that must end the data collection. Death rates are the ideally wrong measure of senescence, because they cannot be studied in individuals, or even in specified groups. If I measure the mortality rate of a group of a thousand people of age eighty, I cannot measure it for this same group at age ninety, because many of the individuals will be missing. The mortality rate of the remnant ninety-year-olds will not be a fair estimate of what the original group would have had at that age. Those now ninety years old obviously had what it takes to live another ten years after reaching eighty. There is no such assurance for someone who has merely reached eighty. The constant removal of those least able to survive introduces a serious but unknown bias in the use of actuarial statistics to measure senescence. Death rates are an important evolutionary cause of senescence. They also show a clear influence of the rates of senescence that have been evolved. They are not an appropriate measure of those rates.

THE WISDOM AND STUPIDITY OF THE BODY
......

Two influential books with rather contrary titles were published earlier in this century: *The Wisdom of the Body*, by Walter B. Cannon (1932), and *Man, The Mechanical Misfit*, by G. H. Estabrooks (1941). Both titles were justified by their contents and both made valid points. The *wisdom* is embodied in the sublime sophistication of our adaptations in general and our cybernetic regulatory systems in particular. It is these phenomena that Cannon discussed and that I emphasized in early chapters of this book. Estabrooks, by contrast, identified flaws in our makeup, many of them related to mechanical imperfections in our upright posture and bipedal locomotion. We are a horizontal, four-legged animal hastily remodeled into a vertical, two-legged one. Estabrooks also emphasized ways in which an animal designed for Stone Age lifestyles may be a misfit in modern ones.

This chapter is more in the spirit of Estabrook's work, and my prime example is senescence. It arises from biologically adaptive trade-offs but surely, from a human perspective, the steady deterioration of our adaptations can hardly be said to represent wisdom. An eighty-year-old really can be a misfit, from evolution having given an enormously higher priority to much younger individuals. Unfortunately, senescence, and the biological conflicts discussed in chapter 6, are not the only aspects of human nature that operate against human values. There are many dysfunctional aspects of the basic design of the human body. By *basic,* I mean those we share with all mammals, if not all vertebrates, not just minor mechanical adjustments required by bipedalism. Other examples, which can be blamed for serious medical problems, will be mentioned in chapter 8. Here I will emphasize unfortunate functional limitations that may cause inconvenience, but not what we would call diseases.

Such problems arise from what is termed *phylogenetic constraint.* Evolution never designs anything from scratch. It can only tinker with whatever happens to be already there, saving those slight modifications that provide immediate benefits, culling those that cause harm. Much of anatomical human nature derives not from anything currently desirable but from adaptive changes made in the early history of the vertebrates. Some of these now underlie functional limitations in ourselves and in vertebrates generally. Many of the most obvious such features date back to the initial establishment of bilateral symmetry. Vertebrates and most other multicellular animals have fore-and-aft, dorso-ventral, and left-right anatomical dimensions. This means that many anatomical structures are either single midline organs, or they come in pairs, one member to the left, another to the right.

For an easily appreciated example, note that we each have one breastbone and two collarbones. Midline structures like the breastbone, or the individual pieces that form it, may be serially repeated any number of times in different species. The vertebrae themselves are a good example, ranging from a dozen to hundreds. Paired features are obviously restricted to even numbers. We have two eyes, four limbs, ten fingers, twenty-six ribs (thirteen pairs). The numbers of fingers and especially of ribs can change in evolution, and this is generally true of parts found in moderately large numbers, such as fingers and ribs. An extra finger or pair of ribs need not cause a major disruption of development. Reduction in size of a finger or rib, a change that could lead to its gradual elimination, would be even easier to evolve.

But suppose it happened that the way of life of some vertebrate would make six the optimum number of limbs. Too bad for it—there is no way some slight modification of the development of a four-limbed animal could provide a basis for evolving an extra pair of limbs. Of course, major modifications in individual development do happen. Two-headed babies are sometimes born to women or ewes or other vertebrates. This kind of major change is always so disruptive as to be immediately lethal. If some mutation did cause a major modification, with a duplication of the shoulder and foreleg structures so that a six-limbed animal was actually brought into existence, and could miraculously survive into adulthood, what would be its chances? Would it be able to compete in a herd of animals with a four-limbed lifestyle? Would it win a mate? Even if it did, would it pass on its six-limbed condition to its two-parented offspring in a functionally adequate form? Surely only discouraging answers can be offered to all such questions.

But really, you might think, since terrestrial vertebrates normally have two pairs of limbs (and most fish two sets of paired fins), there must be something functionally favorable about having two pairs. Wrong! Being functionally favorable or unfavorable for a basic body plan is not a reason for retaining or getting rid of some feature. We have two pairs of limbs not for functional reasons but for purely historical ones: the first lungfish that crawled from the water and pushed its way through the mud did so with the help of two pairs of appendages, and for that reason its descendants still have two pairs. Good or bad, it is a condition those descendants are stuck with, and all they can do is make the most of it.

Is being restricted to four limbs really a serious handicap? Artists in depicting angels have certainly thought so. Angels always have large birdlike wings, in addition to seemingly human arms and legs. Their creators postulated six-limbed vertebrates because angels without arms and hands could not well fill their nuncial roles. In evolving wings, birds seriously constrained their evolutionary futures. They could never again use their forelimbs for terrestrial or arboreal locomotion or manipulation. Hawks and owls must grab and hold prey with their hind feet.

Bird wings, once evolved, do not evolve into anything else. It has never happened that a bird modified its wings for some purpose other than flying, and continued this modification until it lost the power of flight, but could use its wings to carry out the new function to an advanced degree. Penguins might be cited as doing just this, but swimming with wings is mechanically a form of flying, in an aquatic rather than aerial medium. Other birds adopted lifestyles that led to the loss of flight, but this always meant the reduction or near loss of wings. Kiwis are two-limbed vertebrates that gradually evolved from ancestral four-limbed vertebrates.

This line of reasoning suggests that insects have a major evolutionary advantage over vertebrates. A typical insect has six legs and four wings. Evolution can and often has changed insect wings and limbs into something other than locomotor appendages. A mantis uses its forelimbs as special weapons for grabbing prey, but its remaining two pairs of legs still serve admirably for walking. Beetles have converted the first pair of wings into protective armor, but may still fly quite well with the second pair.

Unfortunate numbers-of-parts constraints on human life are perhaps clearest for our paired organs of sight and hearing. Would having some number of eyes other than two be functionally effective? Two eyes are surely better than one, and when pointed in the same direction, as ours are, they allow stereoscopic vision and seeing things in depth rather than less informatively as flat images. Many animals, for instance rabbits, sacrifice this advantage and have the eyes pointed in nearly opposite directions from opposite sides of the head. This gives them the ability to see predators approaching from many different angles, but they thereby lose information about the depth dimension.

Wouldn't we be better off with the two eyes we now have plus a third that would tell us what is sneaking up behind? We could drive without needing a rear-view mirror. With maybe six eyes, we could have precise stereoscopic vision in all directions at once, including straight up. A six-eyed Newton might have dodged that apple and bequeathed us some levity rather than gravity. Constant, full visual information in all directions would enormously enrich our lives and contribute in many ways to safety and well-being. Regrettably, we are stuck with two eyes.

Likewise with ears. Two are better than one, because they help us tell what direction a sound comes from. We do this by comparing the loudness and the phase difference between our two ears. This works well only for horizontal directions. We can easily tell whether a noise comes from left or right, but we are not as good at distinguishing above from below. Sensory physiologists are in fact somewhat mystified that we have any vertical discrimination at all. Apparently we learn to use the different echo patterns from the complex folds of our external ears and the differences in echoes from the ground or nearby objects for clues to the vertical direction of a sound.

For perceiving the distance to a sound source, we have even less adequate evidence. A sound that has traveled a long way before reaching our ears tends to sound a bit different, not merely less loud, because it tends to lose some of its higher-pitched overtones over a long distance. This factor makes a honking car a kilometer away sound different from one 100 meters away, but is of little help in distinguishing a human voice a meter away from one at two meters.

All these limitations would disappear if we had more ears. One on top of the head would enable us to perceive vertical directions as accurately as the horizontal. Three ears would also permit precise distance perception at short range. Differences in arrival times of peaks and troughs of sound waves would tell us not only the direction of the sound but the distance to the source, because these timings would be influenced by the curvature of the sound waves. The diagram on page 130 shows how three ears at the apexes of an equilateral triangle could compare times of arrival of sounds coming from the same direction, to determine how far away their sources are.

So human nature is limited by phylogenetic constraint in its sound-processing capabilities, but what of human artifice? Is there no way for technology to come to our rescue and enable us to perceive distance and vertical direction? Probably not, if by this we mean something as sophisticated as our brain programming for telling the horizontal angle of a sound source. There is more hope for other sorts of improvements in our use of sounds. Hearing-aid manufacturers have accomplished marvels of miniaturization for microphones and speakers and of selective amplification of different pitches. They have inexplicably ignored the possibility of using more than the two microphones mounted in those two hearing aids stuffed into two human ears. They have done nothing to make use of the principle noted in the figure.

What could be done? Marvelous things, that's what. Imagine you are in a noisy pub wearing seemingly conventional hearing aids and, secretly, other things. There are microphones not only in your ears but in the nosepiece of your eyeglasses. The positioning of these sensors would not be much different from those in the figure. Ideally, a fourth microphone under your collar would provide a tetrahedronal pattern that could hear three-dimensionally. Wires convey the sonic information from all sensors to a computer in your shirt pocket. You manipulate dials on that computer and tell

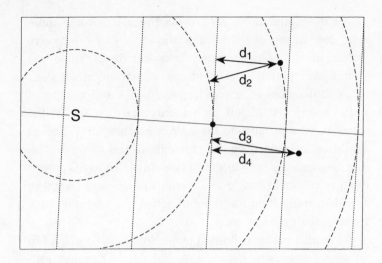

Ears are assumed to be positioned at the •-marks. The dotted lines show locations of peak compression for sound waves originating so far to the left that they have no detectable curvature. The dashed lines represent sound waves originating at the source labeled S, and have marked curvature in the region of the three ears. The difference between the curved and the flat sound waves alters the lag time (proportional to the d's) between perception by the closer and more distant ears. This difference in lag time could be interpreted as indicating the distances to the sound sources.

it: suppress all sounds from less than three or more than five meters away and from any direction other than a narrow sector to my right; amplify all sounds from intermediate distances to the right. The person across the table is regaling you loudly and tediously, but that voice is just a faint murmur. You pretend to heed the regaling but are really eavesdropping on that clever and scandalous conversation at the table four meters to your right.

My free advice to hearing-aid manufacturers: go on trying to help those whose hearing is of less than standard adequacy, but there is much more you ought to be doing. You should use modern technology to help those thought to have good hearing (by the deplorable two-eared standard). Equip them with multiple sensors and data processors to enable them to hear selectively by distance and direction. Nature, in restricting us to two ears, deprived us of much of what we might gain from a sense of hearing, but, as Katharine Hepburn said to Humphrey Bogart in *The African Queen*, "Nature, Mr. Alnutt, is what we were put here to rise above."

··

CHAPTER 8

Most serious illnesses of the more fortunate people in modern societies are degenerative conditions of old age, especially expensive and often fatal cancers and cardiovascular impairments. There are also many other old-age afflictions that, while not likely (nowadays) to be fatal, are the source of much misery and medical expense, such as arthritis, osteoporosis, sexual dysfunction, and impairments of sight and hearing. These geriatric problems now loom so large only because of the abnormally low mortality rates we enjoy in youth and middle age. The miseries of old age are the price we pay for not being killed in childhood or maturity by lions or lungworms.

WHY ARE WE MOSTLY SO HEALTHY?
······

Cannon's classic work *The Wisdom of the Body,* mentioned in chapter 7, proposed that the body's workings represent a suite of superbly engineered adaptations. The engineering *wisdom* shows itself not just in what is there but even more in the machinery that prevents harmful changes that could easily happen. Our tissues are kept at a nearly neutral pH, whether we are living largely on pickled herring and tart fruits, which should lower the pH into the acid range, or on high-calcium foods, which should raise it. Likewise, we usually manage to keep our innards close to 37°C, whether we are working in the sun on a hot day or coping with a blizzard.

The same story can be told of the concentrations of individual substances, whether simple inorganic ions like potassium or sulfate or complex organics such as plasma proteins and hormones. The *wisdom* consists mainly of regulatory processes in the form of sensors of crucial variables such as temperature and prolactin concentration, and the signals from these sensors to the machinery by which the concentration is raised or lowered. Even a slight defect in this regulatory system, with a resulting nonoptimal value of some quantitative variable, can have seriously negative effects on health. Maintenance of life requires elaborate activities by masterful mechanisms.

DESIGN FLAWS THAT LEAD TO ILLNESS
·······

Regrettably, the wisdom of the body is not the whole story. We have to attend not only to the body's impressive cleverness but also to the stupidities that arise from its being the product of natural selection rather than any rational planning. As noted in the last chapter, we share some of these dysfunctional conditions with all vertebrates, or all mammals. Others are uniquely human misfortunes, or affect only those in socioeconomically modern societies, for which evolution has not adapted us. Chapter 7 dealt with what I assumed would be considered minor handicaps and inconveniences rather than serious medical difficulties, like having only two eyes and ears (although some might think the pain and difficulty of childbirth are worse than a mere inconvenience). This chapter will discuss evolutionary legacies that make us vulnerable to serious medical difficulties, and then will take up other Darwinian insights into medical problems.

We are in fact plagued with dysfunctional design features from head to toe, some resulting from evolutionary changes that may have been quite adaptive when they first occurred, often in the early stages of vertebrate or mammalian evolution. For instance, the possibility of choking on food was established long ago with a minute aquatic ancestor's initial invention of a respiratory system. That ancestor was already taking water through the mouth into the forward part of the digestive system and expelling it through a sieving apparatus as a way of obtaining food. In its early stages this ancestor was no doubt small enough that passive diffusion between its tissues and the surrounding water satisfied all its respiratory needs. As some of the members of this group of animals evolved to a larger size, the food sieve, or part of it, could easily assume the additional task of gaseous exchange with the water being channeled through.

This stage, still found in the closest invertebrate relatives of vertebrates, is shown in the top diagram on page 135. All vertebrates have retained this original association of the respiratory and digestive systems. All vertebrates, from fish to mammal, are capable of choking on food. Compared to overeating or reckless driving, this is a minor medical hazard, but it kills some thousands of people every year. Other thousands are saved by the first-aid intervention of the Heimlich maneuver.

At the lungfish stage of our evolution (lower figure), air was taken into the mouth at the water surface through nostrils located, not surprisingly, on the top of the snout. From there it went back past the gills into the paired passages that led from the underside of the pharynx to the lungs. An unfortunate further development in the line leading to mammals was the confluence of these passages into a single

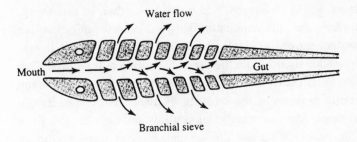

Water flow

Mouth

Gut

Branchial sieve

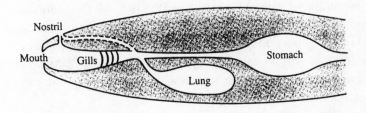

Nostril

Mouth

Gills

Lung

Stomach

Top. Microscopic, wormlike stage of vertebrate evolution as seen from above through overlying tissues. Respiration was adequately served by passive diffusion of dissolved gases between the animal and the water in which it lived. When larger size necessitated a respiratory system, maintenance of the current of water through the food sieve of the pharyngeal region sufficed for both feeding and respiration.

Bottom. The lungfish stage of human ancestry, seen in vertical section to one side of the midline. The dotted connection between nostril and trachea shows the evolutionary restriction of the shared passage to a crossing in the pharynx, as in the line leading to mammals.

trachea, which then branched into the bronchi leading to the separate lungs. This change may have been adaptive at the time, but it assured that the air passage and digestive systems would always have to cross each other. In this evolutionary line, the connection between nostrils and digestive system moved gradually back into the throat until it was immediately opposite the opening leading to the lungs. This is a functionally effective arrangement that almost solves the traffic problem at the crossing of digestive and respiratory systems. Most mammals can swallow and breathe with little interference, because the air passage can form a kind of bridge over the food passage.

We are not up to the general mammalian standard in this respect. Compromises were imposed on this region by the evolution of speech capability. Watch a newborn infant nurse, and you will see no indication of interference with breathing, but by the time it reaches the age of complex babblings, it will occasionally sputter and cough as it tries to swallow and breathe. It is soon as vulnerable as you are to suffocating as a result of eating.

Other medical problems arise from another nonsensical association between systems, the reproductive and the excretory. An early vertebrate ancestor was small enough that passive diffusion was as sufficient for the excretion of wastes as it was for respiration. As our ancestor evolved to a larger size (or left the sea for freshwater), the need for an excretory system gradually arose. Instead of evolving a new tube for the new need of getting excess water and metabolic wastes out of the body, it made use of some plumbing already in place, that which had been serving to get eggs or sperm out of the body. This no doubt worked fine, at first. The urine of freshwater animals is mostly just water, and toxic wastes are so diluted that they should not interfere with reproduction. But consider the mammalian female reproductive system. If it retained the primitive vertebrate arrangement, the kidneys would drain into the oviducts and flow out through the uterus and vagina. The fetus, which has excretory problems of its own, would be bathed in maternal urine, not a wholesome arrangement.

In the course of evolution from fish to mammal, this problem was solved by the connection between the two systems gradually moving farther back. In most mammals today, the urethra discharges into the terminal part of the vagina, where no interference with reproduction can be expected. Women, and their closest relatives among the apes, have actually lost the connection. The urethral orifice has moved outside the vagina. The two systems are now completely separate, although the close association of their openings reflects the recency of the separation.

Note the difference between the problem of minimizing traffic problems where the respiratory and digestive systems cross in the throat, and minimizing interference between excretion and female reproduction. All the excretory and reproductive systems have to do is reach the outside. There is no reason why the shared passage cannot just get gradually shorter until it is no longer there. Perhaps there was once a stage in which excretion made use of the terminal 90 percent of the reproductive system. A few million years later it may have been making use of only the terminal 80 percent. There was nothing in the geometry of the situation to prevent the ultimate elimination of the shared passage. By contrast, once the passage from the nostrils reached a point opposite the opening of the trachea (see lower figure on page 135), no further progression of slight changes could accomplish anything more.

Male mammals clearly lag behind the females in their separation of reproduction and excretion. The urethral passage from inside the lower abdomen out through the penis must serve both functions. Semen must go through a passage where there may be some residue of urine, but I am not aware that this slight contamination causes any difficulty. Still, the arrangement makes no sense. There is no currently adaptive reason for the association of reproductive and excretory systems.

A common medical problem does in fact arise from this association. Many a man in his fifties or later has learned to his sorrow that the reproductive system can cause a problem with the excretory. The prostate gland, which supplies important ingredients to semen, plays an essential role in male reproduction. It is also prone to maladaptive growths, often malignant, in older men. An enlarged prostate presses against the bladder so that it can no longer hold a normal volume, and urination is impeded by mechanical pressures of the prostate on the closely associated part of the urethra. So what logically ought to be just a reproductive malfunction is also a urinary problem, because of the historically determined but currently senseless association of the two systems.

While on the subject, I will mention another functional absurdity in the male reproductive system, even though I am not aware of its causing medical problems. The testicles have moved, in the course of evolution, as in the development of each individual, from deep inside the body to the scrotal sac behind the penis. A current functional reason for the relocation is that it permits sperm production at a degree or two cooler than in the abdomen. This seems necessary for male fertility in most mammals, for reasons not entirely clear. In many species that reproduce seasonally, the testicles enter the scrotum only in the breeding season. When the season ends, they are retracted to a less vulnerable location inside the abdomen.

As the testicles moved ever closer to the point at which they drain into the urethra, it might reasonably be expected that ever shorter tubes would be needed to convey the semen to its destination—that is, if natural selection were a reasonable process. Instead, it is concerned only with what is slightly more adaptive now, and is utterly blind to future consequences of current change. An analogy to what hap-

The gardener's problem seems easily solved: all he need do is go back around the tree to enable the hose to reach the rest of his garden. A much less sensible solution would be to lengthen the hose. This is exactly what happened in the evolution of the male urinogenital system (see figure on page 140).

pened in the evolution of the mammalian reproductive system can be seen in the behavior of the gardener shown in the above picture. He has watered his perennial border to the right and along the back fence, and now wants to continue to the left. Unfortunately the garden hose has caught on the tree. All he has to do is carry the nozzle clockwise around the tree, come back to his present position, and continue watering. But suppose he seeks an additional segment of hose to add to the length he already has. Stupid? Indeed.

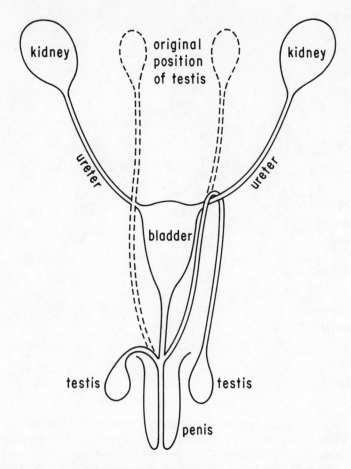

The testis (on the right) draped over the ureter shows the present human male confirguration. It is the result of the gradual movement (in evolution and individual development) of the testicles from deep inside the body (the dashed lines) to the scrotum. The testis on the left shows what might have been if evolution could anticipate future needs or correct past mistakes.

Yet this is precisely the mistake that evolution made in moving the testicles from a central abdominal to the scrotal position. The tube from testicle to the urethra gets hung up on the ureter, which conveys urine from kidneys to bladder. In fact, the gardener is more excusable than evolution. He had to go to the far side of the tree to water his plantings there, and it is understandable that this would have led him to his present position. The testicles, in their evolutionary progression from fore to aft, had no functional reason to go below the ureter—they just did it, as shown in the figure on page 140. Thereafter, the continued movement toward a junction with the urethra required longer rather than shorter tubing.

MEDICAL PROBLEMS VERSUS BIOLOGICAL SOLUTIONS
·······

If something harms you—a mechanical stress, a toxin, millions of *Streptococcus* cells feasting in your throat—you are likely to be aware of it. The harm results in symptoms, such as impairments arising from the damage, and from your own efforts at repair and prevention of further damage. The impairments and repair are evidence that the damage has occurred and may still be occurring. As long as the symptoms continue, you know that you are not yet cured. So the disappearance of symptoms can be good news, but the absence of symptoms does not always mean the absence of damage. A steel blade cutting into your abdomen would normally cause severe pain, an adaptation that would lead you to do whatever may be needed to stop the cutting. Modern anesthesia can completely remove the pain, a fact for which many an appendectomy patient is deeply grateful.

By implication I have been discussing medically good and medically bad symptoms. The surgical wound in the appendicitis patient's abdomen is bad, but necessary if the greater evil of a ruptured appendix is to be avoided. The pain of the wounding would be biologically adaptive in all historically normal situations in which such wounding could occur. It is medically maladaptive in the grossly abnormal event of antiseptic surgery by a skilled physician. The post-operative pain is adaptive to the extent that it leads the patient to avoid additional stress to the injured tissues. It is probably more extreme than is needed, if the patient stays in bed and is well informed and conscientious about following medical instructions. So a postoperative analgesic is surely permissible and may be medically desirable if it permits restful sleep.

So far, the good and the bad may be rather obvious, but consider that *Streptococcus* infection. Symptoms may be headache, painful throat, fever, anemia, and the impaired speech of laryngitis. Should any items on this list be considered good symptoms? Yes, all of them, except the speech impairment. The headache makes you want to relax and avoid stress, a good idea for hastening recovery from an illness. The sore throat means you will not try to shout or talk too much, and will be careful about what you swallow. The fever and anemia are things you are doing to the bacteria, not what they are doing to you. The higher temperature hastens and facilitates immunological response, and the anemia deprives the bacteria of a needed nutrient. The seemingly anemic patient probably has all the iron needed for essential processes, but has removed much of what would normally be in circulation and sequestered it in the liver, where the *Streptococcus* cells cannot get at it.

The table is based on a formal symptom classification proposed in 1980 by Paul W. Ewald, a pioneer of Darwinian medicine. Obviously some such classification is needed for deciding on a treatment for your infection. Surely the distinction between your defenses and your impairments is important, if you want your doctor to help you and not help the strep infection. Just as surely, I regret to point out, doctors seldom appreciate the distinction and its importance, and their medical schools never taught any such material as that in the table. So if a blood test shows that your iron levels are low, your doctor may prescribe an iron supplement, just what the strep needs to help it overcome your defenses. The medical profession gives little thought to what is adaptive or maladaptive for patient or pathogen. Instead it is obsessed with a naive concept of *normalcy.* If something about you is *abnormal,* it should be corrected so that you will be *normal.*

A brief parable may help show why the normal-abnormal distinction is so fallible. You have left work at the end of the day and are approaching your house, when you notice two abnormal things about it. Great clouds of smoke are escaping through the windows, and out front is a big red truck and some men squirting water on your house. Do you immediately try to correct these two abnormalities and start fighting both fire and firemen? Most likely not. You realize that the two abnormalities, like impaired speech and iron sequestration, have different adaptive significance. It is adaptive for you to help fight the fire if you can. It would certainly be maladaptive to attack those firemen.

When you have an infection of any kind, you are engaged in a contest with pathogens, like that between firefighters and fire. You are trying to win, to defeat those enemies, but they are also trying to win. It is imperative to know whether a given symptom results from what you are doing to them or what they are doing to you. A drug that suppresses some symptom that you find unpleasant, for instance, fever, could be seriously counterproductive if your goal is to get well as quickly as you can. This does not mean that fever should always be allowed to rise to its natural level. The human body is perfectly capable of misjudgment in its use of defensive tactics. The fever may often not rise as high as it ideally should for fighting an infection, or it can rise too high and produce serious disturbances such as delirium or tissue damage. Nevertheless, it is probably a good rule not to use drugs for moderate fevers. A *normal* temperature may well be harmful to the victim of an infection. A more general rule is to try to understand the symptoms, for which the table may be of help.

A Classification of Phenomena Associated with Infectious Disease

General Category	Examples	Beneficiary
1. Damage to host tissues	Tooth decay, harm to kidney in nephritis	neither
2. Host impairment	Ineffective chewing, reduced detoxification	neither
3. Repair of damage	Regenerating tissues	host
4. Compensation for damage	Chew on the other side	host
5. Hygienic measures	Slap mosquito, avoid sick neighbors	host
6. Host defense	Coughing, vomiting, antibody production	host
7. Evasion of host defenses	Move away from immune cells, molecular mimicry	host
8. Attack host defenses	HIV destruction of immune cells	pathogen
9. Uptake of nutrients from host	Growth and reproduction of bacteria	pathogen
10. Pathogen dispersal	Mosquito transfers germ to new host	pathogen
11. Pathogen manipulation of host	Exaggerated sneezing or diarrhea, behavior change from rabies	pathogen

The table is, I think, medically wise, but it can be considered flawed from a public-health perspective. Coughing and sneezing, at least at moderate frequency, may be beneficial for someone with a head cold. They help clear pathogens and cellular detritus from the throat and nasal passages. Unfortunately, a pathogen may evolve ways of putting such host defenses to its own use. They can serve to disperse its propagules to a new host, a currently healthy but vulnerable individual. It would not be surprising to find bacterial secretions that efficiently stimulate coughing and sneezing so as to make the current victim cough or sneeze at more than its own optimum rate. This could cost the victim little and greatly benefit the pathogen, at serious cost to anyone nearby.

My discussions in this book have depended heavily on the past, the prolonged history that made us and the rest of the living world what we are today. I have assumed that human nature and our natural environment are the end result of a long but slow historical development. I have implied that evolution is so slow that we cannot effectively watch it happen. There is one outstanding exception to the rule that our natural environment is the result of an extremely time-consuming process. That environment includes microorganisms, and they can evolve with great rapidity compared to ourselves, a rapidity that actually enables us to watch it happen. This rapid evolution gives microorganisms an immense advantage in their arms races with us.

It is now widely realized that use of antibiotics for fighting bacteria is not as effective as it was some years ago. The bacteria have been evolving resistance. With many generations per day, they have time for an appreciable amount of evolutionary change in a few years, sometimes even in a few weeks. A bacterium can have perhaps two thousand generations in a month. How long would two thousand human generations take? It is not really clear how great a problem antibiotic resistance will be in the future, but the outlook is alarming. Diseases easily cured by penicillin and other antibiotics twenty years ago are not as easily cured today by these or even more recently developed drugs. This tragedy is largely of our own making, from the routine use of antibiotics on livestock and for inappropriate medical treatments. This problem is now getting close attention from medical researchers and public-health activists.

It is not as widely appreciated that microorganisms can rapidly evolve in other medically important ways. Their virulence, for example, can change rapidly in relation to changed conditions inadvertently imposed by human activities. Serious pathogens can evolve to be less serious, and vice versa. The evolution of virulence is a complex process controlled by many factors, but one of the more important ones is perhaps readily appreciated. If different species of parasite, or genetically different strains of the same species, are competing within a host, the most virulent parasite or strain is most likely to win. Each competitor is exploiting the host to make more of itself, including infectious propagules to be dispersed, one way or another, to new hosts. The one that most vigorously exploits the host will thereby be reproducing itself more successfully than the others. If the harm it causes is so great that the host dies, this deprives that greedy parasite of its livelihood. But it harms the competitors just as much, and they did not enjoy the prior bonanza. So natural selection among competing parasites within a host tends to favor the most virulent.

But another kind of selection is surely taking place. Suppose there are two infected hosts. One has pathogens that let it live a long time, and the other those that kill it quickly. The pathogens in the long-lived host will be dispersing propagules to other hosts for a longer time than those in the host that dies quickly. As a group, those more benign parasites in the long-lived host are more successful. So there are two levels of selection taking place: among pathogens within a host and among groups of pathogens in different hosts. The levels of virulence evolved will reflect (among other things) the balance between this within-host and between-host selection. Many environmental and other factors will affect this balance.

With reduced virulence as the medically desirable condition, anything we can do to reduce selection within or intensify it between hosts may have the desired effect. Any human activity that reduces the rate of spread of a pathogen from one host to another ought to intensify selection against virulence. If transmission to new hosts is rapid and frequent, even a really virulent pathogen may infect many new hosts before the original one dies. This death would be just a minor loss to the pathogen. By contrast, if transmission is sufficiently slow and infrequent, the first host's death may occur before any new ones are infected, and the pathogen would have caused its own destruction by its high virulence.

So whatever impedes the spread of a pathogen from one host to another should be desirable, not just as a public-health goal in itself but as a way of selecting for reduced virulence in the individuals that do contract the disease. General use of condoms and clean needles should result in less rapidly lethal HIV; window screens and mosquito repellent should select for less virulent malarial parasites; sanitary water supplies should cause cholera to be less of a scourge to those who catch it. I anticipate that the evolution of microbial virulence, drug resistance, and other traits of medical or public-health interest will be the focus of much research and activism in the coming years.

The infectious-disease table on page 144 is relevant to medical problems other than infectious disease, for instance, to poisoning or injuries of any kind. With these nonliving sources of distress the problems are much simplified, in that no benefits for some enemy nor possibilities of evolutionary change need be considered (for example, for drug overdose or sunburn). So the lines in the table for symptoms that benefit the pathogen can be eliminated. The important point is that symptoms indicative of harm should always be distinguished from those indicative of the victim's efforts at recovery.

ABNORMAL ENVIRONMENTAL FACTORS
......

Of the many other ways the ideas of evolutionary biology are relevant to medical problems, I will discuss just one more here. The process of natural selection is inescapable. It is no doubt as operative today in human populations as it ever has been, but it is a slow process by most human standards. I have assumed in this book that we have had time to evolve very little since the invention of agriculture a few millennia ago. Human nature is now very much what it was in the Stone Age, which lasted more than a hundred times as long as recorded history. Current human nature is designed for Stone Age life.

But the human environment has changed radically in the last few thousand years. Agriculture caused a drastic shift in lifestyle. It suddenly transformed our diets and allowed populations to expand enormously. Permanently settled villages and the later big cities can be supported only with abundant farm products. Farming led to the explosive development of technologies that are growing faster today than ever. Most of the technological developments have been beneficial, leading to the many luxuries and securities we now enjoy. A comparison of the personal freedoms and good health and low mortality rates of modern populations with those of any previous time shows that our world is marvelously blessed.

But it is not perfect. The sweetness of the health and safety we enjoy for most of our long lives is paid for by the bitterness of old-age impairments and miseries. Other benefits also have their costs to our health. The abundant products of agriculture do not provide ideal diets. Flour and the milk of other animals allow for early weaning. This can augment fertility, but can have adverse effects on children's health. Diets with abundant starch and fat and protein for children and adults could be deficient in the vitamins that were normally abundant in ancestral hunter-gatherer foods. In the archaeological record, the transition from foraging to farming is sometimes documented in skeletal remains by the farmers' smaller stature, frequent bone defects, and dental pathologies.

In modern industrial societies, it may be that the main causes of illness are the mismatches between our Stone Age adaptations and our modern environments. A prime example is the problem caused by our dietary predelictions and the foods readily available to anyone browsing supermarket shelves or a restaurant menu. In the Stone Age there was a consistent advantage in going after foods that were as sweet and tender and rich as could be found. This led people to avoid the potent chemical weaponry of most plants by seeking ripe fruits and bland tubers and the most easily eaten parts of whatever wild animals could be hunted. These were most likely such things as lizards and snakes and insects. The technology of hunting sizable mammals and birds (such as archery and the domestication of dogs) arose late in the Stone Age and was often a seasonal luxury. Maximizing intake of sugar and fat normally led to health and vigor. Salt was also an essential nutrient often in short supply.

We have the same Stone Age motivations today, but have easy access to many times the historically normal levels of sugars and fats and salt. The result is undoubtedly a much higher incidence of obesity, diabetes, cardiovascular disorders, and many kinds of cancer than we would have on normal Stone Age diets. A related problem is our habitual physical inactivity. We make our livings sitting at desks or assembly lines or behind steering wheels rather than dashing about in the fields or laboriously digging roots or climbing or stooping for fruits. This sedentary life satisfies our urge to save energy, an urge of great value in the Stone Age, but now a liability when combined with our excessive caloric intake.

The conflict between our feeding adaptations and environmental largess is perhaps the most easily appreciated of the many mismatches between our evolved adaptations and current environment. There are many others. A good rule, for thinking about this sort of problem, is to be alert to the prevalence of maladies that would be seriously disadvantageous in a normal (Stone Age) environment. The appropriate question then is: Why has natural selection not eliminated our vulnerability to these conditions? One possibility is that the vulnerability is the price paid for some beneficial effect, perhaps in other individuals or another part of the life history. Senescence as a price paid for part of our youthful vigor would be a good example. Resistance to malaria is a benefit for many individuals from the gene that causes sickle-cell anemia in a few.

Another possibility is the one emphasized here, that the problem arises from a mismatch between Stone Age adaptation and some aspect of the modern environment. Good examples are myopia and the need for such services as orthodontia and wisdom-tooth surgery. Imagine the plight of a forager on the African savanna who can't tell a rock from a rabbit or distinguish a friend from a deadly enemy at the distance a spear might be thrown. Or endow this same individual with misplaced incisors and painfully impacted wisdom teeth. Myopia and these dental problems are seldom found in today's tribal societies, and this must have been true in the Stone Age. Yet they are common now. Is it not likely that they arise from some abnormal usage of our eyes and of our teeth and jaws during childhood and adolescence? Wouldn't it be a good idea for researchers to find out whether this suggestion is valid, what the abnormal usages might be, and how their effects might be alleviated?

I have barely touched on a few possible ways in which

evolutionary principles may have a crucial bearing on medical education, research, and practice. Many more examples could be listed. I suggest that there is no kind of medical specialty that is not crying out for assistance from Darwinian insights: mental disorders, allergy, cancer, sexual and reproductive dysfunction, insomnia, sunburn, and the whole array of poisonings and injuries. I anticipate that the study of evolution, always of immense intellectual interest, will soon be recognized as an indispensable foundation for medical science.

CHAPTER 9

PHILOSOPHICAL IMPLICATIONS

considering everywhere
Her secret meaning in her deeds,
And finding that of fifty seeds
She often brings but one to bear.

—*Alfred Lord Tennyson*

ALICE: I can't remember things before they happen.
WHITE QUEEN: It's a poor sort of memory that
only works backwards.

—*Lewis Carroll*

In the early chapters of this book I attempted to show that the eighteenth- and nineteenth-century *God-is-smart* tradition is fallacious. Now I plan an attack on what I'll call the *God-is-good* tradition.

Perhaps I should take a moment to deal with what I mean by *God*. I am not an atheist flaunting a caricature to offend people's religious sensitivities. In any theological discussion, I prefer to define atheism out of existence. Whatever entity or complex of entities is responsible for the universe being as we find it, rather than some other way or not there at all, can be called *God*.

With God the Creator so defined, we can proceed to characterize and evaluate Him using the one source of evidence we have: His creation. On that basis I argued, in the first two chapters, that, contrary to the claims of Paley and others of the natural theology school, there is no evidence that God has any engineering expertise. Their arguments failed to recognize that functional design can arise not just from intelligent planning but also from blind trial and error. They failed to recognize that the apparently purposive structures and activities of living organisms are just what we would expect from trial-and-error production. Organisms show the expected stupid mistakes, the dysfunctional design features, that arise when understanding and planning are entirely absent.

The God-is-good idea is also a common supposition, but it cannot possibly be valid if natural selection underlies all functional design. The only thing that anything in nature is designed to accomplish is its own success. Whatever is having greater success now, by whatever means, will have its characteristics more abundantly represented in the future. This is the only kind of reward offered in God's creation. The moral unacceptability of natural selection is not just a conclusion to be asserted or accepted, but one to be thought about. The British literary giant George Bernard Shaw thought about it, and responded with: "when its whole significance dawns on you, your heart sinks into a heap of sand within you. There is a hideous fatalism about it, a ghastly and damnable reduction of beauty and intelligence, of strength and purpose, of honor and aspiration." Shaw's condemnation of natural selection is justified, but the outlook may not be as pessimistic as he seemed to believe. He did not appreciate that, although the biological creation process is indeed evil, it is also abysmally stupid. We can have some hope that our intelligent efforts to circumvent the evil can triumph over so unreasoning an enemy. We can hope, with Thomas Huxley, that "[i]n virtue of his intelligence, the dwarf bends the Titan to his will," or in Richard Dawkins's words, that we can successfully rebel "against the tyranny of the selfish replicators."

With what other than condemnation is a person with any moral sense supposed to respond to a system in which the ultimate purpose in life is to be better than your neighbor at getting genes into future generations, in which those successful genes provide the message that instructs the development of the next generation, in which that message is always "exploit your environment, including your friends and relatives, so as to maximize our (genes') success," in which the closest thing to a golden rule is "don't cheat, unless it is likely to provide a net benefit"?

THE IMMORALITY OF THE PRODUCTS OF NATURAL SELECTION
......

A forest or a coral reef under a blue sky and bright sunshine gives an impression of tranquillity and harmony, but the impression disappears when details are examined. Look at one of those trees in the forest. Almost certainly you will find that it is afflicted with pests and diseases and is under frequent attack by browsers such as deer or howler monkeys. The same can be said of those monkeys. Casual examination of their skins may show the ravages of fleas or ticks or fungi. They live in constant danger from attacks by jaguars and other predators. The story of the forest or coral reef is a tale of relentless arms races, misery, and slaughter.

Back to the tree. It may be producing seeds—hundreds, or perhaps many thousands, at each seasonal episode of reproduction. The tree will ultimately die and be replaced, perhaps by another individual tree like itself. But what of that astronomical number of seeds that our tree produced? Tennyson lamented odds of one in fifty, but that would be utopian compared to the prospects for the seeds of one large tree. Think of all those potential trees that must be listed among life's failures. This is numerically a rather extreme example. The monkey is extreme in the other direction, because a substantial proportion of baby monkeys may survive to adulthood, but even here, the failures outnumber the successes.

Our experience of human life histories today, with the great majority of our babies surviving to adulthood, is grossly abnormal. Human populations, in the historically normal circumstances that prevailed until a few thousand years ago, had what by current standards were high mortality rates. Human populations remained sparse for hundreds of thousands of years. This means a very nearly zero rate of increase for long ages. An increase of 1 percent per century would double the population in seven thousand years. In a hundred thousand years it would make it 22,000 times the original number, but nothing of the sort happened. Conditions of Stone Age life must have been enormously variable from time to time and place to place, but we cannot be far wrong in assuming that adolescent girls would on average live to produce about four children before death or menopause ended their output. This reproductive career would entail many years of lactation, which would inhibit ovulation, delay pregnancy, and keep lifetime fecundity down to a few babies. Only about half would survive to adolescence. Disease, predators, accidents, and murder, often infanticide by hostile tribesmen or even members of their own group, would claim many lives and keep numbers down to perhaps one individual per square kilometer over the more favorable parts of habitable regions.

The infanticide I mentioned is not a social pathology found only in abnormal circumstances. It is prevalent today in diverse human cultures, including some that we might not think of as primitive; it is widespread in a large proportion of the animal kingdom; and it is entirely to be expected from what we know of evolution. These assertions are abundantly documented in the technical literature of anthropology and biology, and infanticide is just one small detail of a monstrous picture. Mountains of data on parasitism and predation (including cannibalism) in nature could be amassed to document the enormity of the pain and mayhem that arise from adaptations produced by natural selection. For one example of parasitism, try the heart-wrenching death of Little Echo from meningitis as detailed by Thomas Mann in *Doctor Faustus*. In this chapter I will confine myself to the specific example of infanticide described by the California anthropologist Sarah Hrdy.

She studied a population of monkeys, Hanuman langurs, in northern India. Their mating system is what biologists call *harem polygyny*: dominant males have exclusive sexual access to a group of adult females, as long as they can keep other males away. Sooner or later, a stronger male usurps the harem and the defeated one must join the ranks of celibate outcasts. The new male shows his love for his new wives by trying to kill their unweaned infants. For each successful killing, a mother soon stops lactating and goes into estrous. The death of her infant converts her more quickly from a potential to an actual resource for the male's reproduction. This is why infanticide is adaptive for the male.

His murderous efforts do not always succeed. The females are often sisters or other close relatives and may share a genetic interest in the survival of a threatened baby. So the mother may have help in defending her offspring. Unfortunately the male is much bigger and stronger and often does succeed. Deprived of her nursing baby, a female soon starts ovulating. She accepts the sexual advances of her baby's murderer, and he becomes the father of her next child.

Do you still think God is good?

Awareness of the prevalent wickedness of what had been personified as Mother Nature is a recent development. Theorists and field workers have understandably preferred to study and discuss the pleasanter aspects of human and animal behavior. My favorite example of what might be called willful ignorance by a field biologist is from a 1916 study of Antarctic penguins by C. Murray Lavick: "Many of the colonies are plagued by little knots of 'hooligans' who hang around their outskirts, and should a chick go astray it stands a good chance of losing its life at their hands. The crimes that they commit are such as to find no place in this book." Unlike many later workers, Lavick was honest about his censorship.

Hrdy was the pioneer in bringing the prevalence of infanticide to the attention of biologists, social scientists, and the scientifically literate public. Her 1977 article "Infanticide as a Primate Reproductive Strategy" was greeted with outraged disbelief by many readers who refused to believe that adult males' attacks on infants could be adaptive and normal.

Times have changed, as noted in recollections by the Canadian psychologist Martin Daly at a 1993 meeting in Binghamton, New York. He had attended an address by a biologist who studied conflict over the use of dead mice in a species of carrion beetle. A female bettle lays eggs on a mouse corpse and defends it against other females seeking a corpse on which to lay eggs. A challenger may sometimes appropriate a dead mouse from a guarding female. Someone in the audience asked, at the close of the presentation, about what happens to the young of the first female. The speaker immediately explained that the new female kills them, "of course." Daly regretted Hrdy's absence, imagining that she would have felt a smug satisfaction at realizing that her heretical findings were, less than twenty years later, simply expected.

OTHER MORAL FALLACIES
......

Many traditional religions foster attitudes that ought to have disappeared as biological understanding accumulated over the last century. One of these might be termed the holy-corpse fallacy. When people die, their relatives and friends behave as if there were some moral significance in the dead body. They ignore the fact that the "last remains" are just that, the material that happened, at the time of death, to provide the medium of expression for a human life. However long this complex human message was expressed is the duration of time in which the materials were coming and going, no less than in the candle flame I discussed in chapter 7. The tons of matter that at one time or another were part of a dead senior citizen are already dispersed throughout the terrestrial ecosystem. A small minority of the dead person's molecules are in orbit around the Earth or sun. Cremation of the matter that happened to be there at the last minute merely hastens an inevitable process.

The holy-corpse fallacy once had support from the biological concept of *protoplasm*, the special living matter of an organism. Other matter may be entering and leaving a living cell, but its protoplasm was presumably a stable entity that regulated this material flux. An organism was thought to be a machine, like an automobile, always keeping its resources distinct from itself. A dead person may have dead protoplasm, but it was presumably that person's very own protoplasm, and had been throughout his or her life. Protoplasm was often discussed in the biology courses I took in the 1940s. It is a term almost never heard today.

Many other errors arise from the unjustified idea that human life can have a simple biological definition. This fallacy leads people to express moral objections to such practices as the replacement of human parts with those of some other species. A man whose defective heart has been removed and replaced with one from a pig is somehow felt to be no longer fully human. Biologically he may indeed be 1 percent *Sus* and only 99 percent *Homo*, but if he still has his human hopes and fears and memories, his biology is morally irrelevant.

The moment-of-conception fallacy is a related error. The joining of a human egg and sperm define a new and unique human genotype. It does not produce any human hopes and fears and memories or anything else of moral importance implied by the term *human*. The newly fertilized egg may have the potential for a fully human existence, but that potential was there even before fertilization. The same can be said of all the fertilizations that might have been. The entrance of that egg by one sperm meant an early death for millions of competing sperm. It destroyed all hope for those millions of other unique human genotypes.

Besides being philosophically unacceptable, the moment-of-conception fallacy is biologically naive. It implies that fertilization is a simple process with never a doubt as to whether it has or has not happened. In reality, the "moment" is a matter of some hours of complex activity. There are elaborate biochemical interactions between the sperm and various layers of the egg membrane. The sperm gradually breaks up, and only its nucleus is established in the egg. Then both egg and sperm nuclei initiate radical changes, with condensations and movements of chromosomes, before the fusion of the two nuclei. Many of the developmental events following this fusion were predetermined during the production of the egg. Genes provided by the sperm do not have discernible effects until embryonic development is well under way. A strictly biological definition of humanity would have to specify some point in this elaborate program at which the egg and sperm have suddenly been endowed with a single human life.

There are other difficulties with defining humanity as beginning with fertilization and the establishment of a unique genotype. If that single human life develops for a while and then divides to produce identical twins or triplets, are they, despite their physically separate lives, to be considered one human being? This would be contrary to almost everyone's moral sensibilities. Recent observations have raised additional questions about the connection between biological and moral individuality. Early in development, fraternal twins from two separate fertilized eggs may fuse and develop into what, at birth, is physically a single baby. Molecular techniques available today may show that such an individual is genetically different in various parts of its body. An apparently normal woman may have some genetically male tissues from what originated as her twin brother, or vice versa.

The only realistic view is that a human life arises gradually. A child's acquisition of speech, and use of it to convey ideas to others, is perhaps the most obvious indication of this process. This gradualism is not much help in making personal decisions or devising public policy. We want clear and simple rules as guides for human behavior and the recognition of who should be accorded human rights. The recognition of full humanity in a full-term newborn would be one such simple rule. I am not inclined to argue that it is the best possible, but it makes more sense than any recognition of fetal rights. All the usual arguments for rights before birth are based either on an untenable biological definition of humanity or on fetal behavioral attainments. The observable capabilities of a human fetus can all be matched in other mammals at comparable stages of development.

THE WILSON MANIFESTO
......

The first paragraph of the first chapter of E. O. Wilson's classic work *Sociobiology: The New Synthesis* proposes that a modern biologist realizes that self-knowledge is constrained and shaped by the emotional control centers in the hypothalamus and limbic systems of the brain. These centers flood our consciousness with all the emotions—hate, love, guilt, fear, and others—that are consulted by ethical philosophers who wish to intuit the standards of good and evil. What, we are then compelled to ask, made the hypothalamus and limbic system? They evolved by natural selection. That simple biological statement must be pursued to explain ethics and ethical philosophers, if not epistemology and epistemologists, at all depths.

Wilson's book is a magnificent achievement and probably the most philosophically important biological work of this century, but perfection is an elusive quality. I would fault the book in some minor ways and two major ones related to the above statement. The first is that when Wilson finally (pp. 559–64) discusses the implications of the hypothalamus and limbic systems for ethics and related topics, he fails to follow the lead of his earlier statement far enough or in sufficient detail. My second criticism is that the statement itself is unduly restricted. Why "self-knowledge" and not knowledge in general? And why just the parts of the brain most concerned with emotions? Why not the whole system of neural control and cognition? Why just ethics and epistemology?

More recently, D. S. Wilson (no relation) examined this more general connection between knowledge and evolutionary theory. He pointed out the obvious inference that natural selection, in favoring and preserving sensory and cognitive abilities that contribute to genetic survival, need not be endowing organisms, human or otherwise, with any direct perceptions of things as they really are. It is enough that the perceptions lead to useful responses. Our ancestors seldom moved more than a few days' walk from where they were born, and never looked at celestial objects through telescopes. They perhaps thought of the Earth as a horizontal disk and the heavens as a dome on which the heavenly bodies travel along daily paths. This was an often useful cosmology and never caused any practical difficulties that could relate to survival and the production and rearing of offspring.

Not only the sense organs but also the processing machinery in our brains are there for no other reason than their contributions to genetic success. They were not evolved to enable us to infer what either astronomical or terrestrial objects are really like. Similar limitations apply to all senses and all capabilities for reasoning about ourselves and the world we inhabit. Reasoning, to be favored by selection, must lead to useful conclusions that help us survive and reproduce. It need not lead to formally correct solutions to logical problems.

This insight has been fruitfully exploited by recent psychological researchers, most notably the husband-wife team of Leda Cosmides and John Tooby. Their research shows that the human reasoning process may lead to conclusions that are useful and intuitively correct, even though they must be considered erroneous in formal logic. This will be reliably true when the logically correct solution is less likely to be useful. For instance, it does not follow logically that playing Russian roulette will harm you. So the avoidance of such a game is not logically justified. Needless to say, it is advisable in a practical sense. A more interesting finding concerns problems with formally similar structures that we might think equally easy to solve. Cosmides and Tooby have shown that a problem's ease of solution can be very much a function of logically irrelevant details. People are much better at solving problems with the emotional content provided by certain kinds of human interactions. A problem of guarding against unfair exploitation in social interactions, such as an officemate's pretending to be sick to avoid some unpleasant duty, may be more easily and rapidly solved than what is formally the same problem without the emotional baggage of unfairness.

The inherent limitations of human thought processes are perhaps most serious in their intuitive perception of the passage of time. We take it as intuitively obvious that an absolute present constantly converts an immediate future into a recent past. The physics of the concept of time has lately reached many nonphysicists through discussions in nontechnical books for lay audiences. Physicists currently recognize that time is a basic concept, and that one part of the time scale is different from another, in total entropy, for example. Think of all those equations of the form $y = f(t)$. Physicists find no support in their experiments or calculations for the concept of a present that intervenes between a past and a future. It is as if the universe is simply a historical document, with each successive chapter different in a predictable way from the one before and the one following, but with no bookmark to show where the reading has reached.

This picture is one of extreme determinism. Not only is the future predetermined by what has gone before; it is, in a sense, already there. I am sure this view will be challenged, is in fact being challenged now, and will no doubt be greatly altered in the future. I also suspect that the future concepts of time provided by physicists, if historical analogy is any guide, may be even less intuitively acceptable than what they have now. I hope there is some basis for expecting some resolution to the problem of the intuitive sensation of the present. In his book *About Time*, Paul Davies seems to expect that "physics will someday explain why we have that inescapable perception of what we call the present and of its movement through that document of history."

Following D. S. Wilson's lead, I can confidently state why we have that intuitive perception: it has been useful in achieving genetic success. I cannot give this idea the sort of formulation it would need to be scientifically useful. By this I mean a theoretical model from which I might predict the results of explicitly described investigations. Like Davies, I hope for some such real progress in explaining the human idea of time, but I think it more likely to come from a biologist, not a physicist. I expect it to come from some young, nimble-minded biologist who knows the physical concept of time and the biological principle of natural selection and can put them together to explain Alice's constraint, stated at the beginning of this chapter.

DOMAIN MIXES
......

The libraries are full of writings in which authors examine nature and extract lessons on how people ought to behave. They always do this with "a lubricious slide from an *is* to an *ought*," in the words of a distinguished historian of science. The philosopher David Hume objected to this practice more than two centuries ago in what is sometimes called "Hume's law," that moral directives cannot be deduced from descriptive premises. Yet no matter how often this law has been confirmed or invoked, it continues to be broken in lengthy printed arguments in which the lubricious slide, though always present, may be elaborately camouflaged.

The lubricious slide that Hume and others have objected to is a special case of a more general kind of fallacy that might be called a domain mix. A *domain* is some aspect of existence, such as the material universe, that can be discussed with a recognized list of descriptors. The appropriate material descriptors include such things as *length* and *mass*. With a few such terms we can define others, such as *density*, something proportional to mass and inversely proportional to the product of three lengths. We can define new terms and also derive physical consequences using these descriptors, for example, the trajectories of objects moving in a gravitational field. We cannot derive consequences in one domain using descriptors peculiar to another (that is, a domain mix). For instance, we cannot deduce moral conclusions from physical premises (Hume's law).

How many domains should be recognized? I tentatively suggest four: the *material*, the *moral*, the *mental*, and the *codical*, as indicated in the chart. I also propose that endless confusion has resulted from people committing domain mixes. The fact that time is an essential descriptor for every domain makes it easy to establish a chronology of events in different domains. A message (codical) may be sent prior to its perception (mental), which is then followed by some action (material). To explain or infer an event in one domain from premises in another requires more than chronology. We need descriptors applicable to the temporally ordered events themselves.

Material	Moral	Mental	Codical
time	time	time	time
mass	good	decision	baud
color	evil	surprise	byte
force	conscience	hope	fidelity
charge	virtue	perception	redundancy
length	vice	knowledge	editing
pH	guilt	belief	text

The codical domain, from the Latin *codex*, meaning "document," is the domain of information. Biological discussions are often confused by an undisciplined mixing of material and codical concepts. The term *gene*, for instance, is sometimes used for a DNA molecule (a material object) and sometimes for the message coded by the base-pair sequence in the molecule. I believe that *gene* is best used only for the message, and that DNA should be recognized as the material medium in which such messages are normally archived. A message can always be expressed in more than one medium, as is apparent for any book printed on paper and also coded as magnetic patterns on an audiotape. Likewise, a gene can be expressed in such media as RNA and protein or, nowadays, as a sequence of the symbols A, C, G, T on paper (as I did in chapter 2). Yet, no matter how diverse the media, it can always be the same message.

A triple domain mix, mixing the material, codical, and mental, has been rampant in recent years. Libraries are ever more burdened with works that start out discussing the physical processes going on in nerve cells, such as charge transfers across membranes and interactions of the molecular structures of hormones with those of nerve endings; soon the authors slip into the codical realm to establish the high sophistication of information processing carried out by nerve cells, but seldom being clear about when they are discussing the medium and when the message. Then, in complex sentences, come the lubricious slides into discussions of pleasure and anxiety and other concepts proper to the mental domain. From such flights of unreason the authors claim to have provided a physical explanation for mental phenomena.

Avoiding domain mixes can be difficult, because such vernacular terms as those in the chart often find uses in more than one domain. A burden is usually something that can be measured in kilograms, and is properly a material concept. But what about that burden of grief I mentioned in chapter 4? This is obviously a metaphorical usage, and nobody would try to measure this kind of burden in kilograms. Unfortunately, the prevalence and general acceptance of such metaphorical expressions facilitate the lubricious slide between domains.

The problem is harder than you might think, because it is not always easy to decide which is the literal and which the metaphorical usage, and I slipped an example into the chart. Guilt can be a feeling (mental) about oneself or a public judgment (codical), perhaps as the verdict of a jury trial, in addition to being something one ought (morally) to avoid. As a matter of personal taste, I think of guilt as most literally a mental phenomenon and only metaphorically useful in the moral and codical domains.

Earlier in this chapter I criticized E. O. Wilson for what I felt to be inadequacies in his use of the theory of natural selection to understand human nature and the current human condition. Now that I have tried my hand at the same job, I am less inclined to be critical. It is an immensely challenging task that no one can hope to bring to anything like a finish. We can only try, and must. Natural selection is a process of pervasive importance in the biological world, which includes our own species, and on which that species is utterly dependent. Progress in evolutionary biology and its applications is perhaps most obviously relevant to medical and environmental issues, but there is no aspect of human life for which an understanding of evolution is not a vital necessity.

••

PREFACE
••••••

vii The term *adaptationist program* was first used, disparagingly, by Stephen Jay Gould and Richard C. Lewontin in their article "The spandrels of San Marco and the panglossian paradigm: A critique of the adaptationist program," in *Proceedings of the Royal Society* (London) B 205 (1979):581–98.

INTRODUCTION: PLAN AND PURPOSE IN NATURE
••••••

3 Schliemann's quest for ancient Troy actually led to more than he bargained for, because his discovery showed there to be several successive occupations of his supposed Troy. It is not clear which occupation, if any, corresponded to King Priam's city, and other nearby localities were also built up at various times. It remains likely that Schliemann's site, or someplace close by, was the source of the Homeric legends, and that Schliemann's reasoning from the available evidence facilitated important archaeological discoveries.

I: ADAPTATIONIST STORYTELLING
••••••

6 The Trees of Valinor are described in chapter 1 of J. R. R. Tolkien's *The Silmarilion* (Boston: Houghton Mifflin, 1977). William Paley's book is *Natural Theology* (London: Charles Knight, 1836). The quoted paragraph is in chapter 3, the watch reference in chapter 1. Paley's theological interpretations, as countered by Darwin's natural selection, inspired the title to Richard Dawkins's *The Blind Watchmaker* (New York and London: Norton, 1986).

11 Galen discusses the human hand in chapter 6 of *The Usefulness of the Parts of the Body*, trans. M. T. May (Ithaca, N.Y.: Cornell University Press, 1968).

12 The antiquity of fishhooks is discussed by Guðni Þorsteinsson in *Veiðar og Veiðarfæri*, (Reykjavík: Almenna Bókafélagið , 1980).

15–17 J. W. Hastings's pony fish article is in *Science* 173 (1971): 1016–17. The quotation from Karl R. Popper is from his article in *Studies in the Philosophy of Biology*, ed. F. J. Ayala and Theodosius Dobzhansky (Berkeley: University of California Press, 1974). The quotation from Ernst Mayr is from p. 328 of "How to carry out the adaptationist Program," *American Naturalist* 121 (1983): 324–34. The work by Stephen Jay Gould is "Sociobiology and the theory of natural selection," in *Sociobiology: Beyond Nature/Nurture?*," ed. G. W. Barlow and J. Silverberg, AAAS Selected Symposium, 35 (1980):257–69. Mark Twain's "A Double-Barreled Detective Story" has often been reprinted, for instance, in *The Complete Short Stories of Mark Twain*, ed. Charles Neider (New York: Doubleday, 1957).

18 Conant's work is *On Understanding Science* (New Haven, Conn.: Yale University Press, 1947).

2: FUNCTIONAL DESIGN AND NATURAL SELECTION
. . . .

21 Hutton's words have been quoted many times, for instance, on p. 560 of Konrad Krauskopf's *Fundamentals of Physical Science* (New York: McGraw-Hill, 1953).

23 The pigeon pictures are based on descriptions and illustrations from several sources, including August Weismann, *Evolution Theory* (London: Arnold, 1904); H. P. Macklin, *A Handbook of Fancy Pigeons* (Palma de Mallorca, Spain: Divers Press, 1954); and Wendell M. Levi, *Encyclopedia of Pigeon Breeds* (Jersey City: T. F. H. Publishers,1965).

24 I can certainly recommend *The Voyage of the Beagle*, Darwin's own account of his years as a naturalist at sea. Like most of his other books, it has been reprinted many times.

26 The picture of Darwin's finches is from p. 19 of the classic work on the topic, *Darwin's Finches*, by David Lack

(Cambridge: Cambridge University Press, 1947). A splendid modern treatment is P. R. Grant, *Ecology and Evolution of Darwin's Finches* (Princeton, N.J.: Princeton University Press, 1986). The topic is also discussed in a superb popular work, *The Beak of the Finch,* by Jonathan Weiner (New York: Knopf, 1994).

27–29 Wallace's work was reprinted on pages 210–27 of *A Delicate Arrangement,* ed. A. C. Brackman (New York: Times Books, 1980). Endler's book was published by Princeton University Press in 1986. Darwin introduced the idea of sexual selection in his classic *The Descent of Man, and Selection in Relation to Sex* (New York: Appleton, 1871). The best modern work on sexual selection is Malte Andersson's *Sexual Selection* (Princeton, N.J.: Princeton University Press, 1994). Botanical applications of the theory of sexual selection are discussed by Mary F. Willson and Nancy Burley in *Mate Choice in Plants* (Princeton, N.J.: Princeton University Press, 1983). The classic work on selection for social status in the animal kingdom is by Mary Jane West-Eberhard, "Sexual selection, social competition and speciation," *Quarterly Review of Biology,* 58 (1983):155–83. The history of Darwinism, especially that of Darwin's ideas on natural selection and sexual selection, is traced in Helena Cronin's carefully researched and splendidly written *The Ant and the Peacock* (New York: Cambridge University Press, 1991). My information on Incititus is from Robert Graves's *I, Claudius* (New York: Knopf, 1934).

31 H. C. Bumpus's work on the sparrows is in "The elimination of the unfit as illustrated by the introduced sparrow," *Biol. Lett. Mar. Biol. Woods Hole* 11 (1896–97): 209–26.

32 The Hume quotation is from p. 297 of *The Essential David Hume* (New York and Toronto: New American Library, 1969). There are several excellent recent works on life-history theory. Two that I would recommend are Stephen C. Stearns, *The Evolution of Life Histories* (New York: Oxford University Press, 1992), and Eric L. Charnov, *Life History Invariants* (New York: Oxford University Press, 1993). Optimal foraging theory is dis-

cussed in many recent books on animal behavior, for instance, John Alcock, *Animal Behavior*, 5th ed. (Sunderland, Mass.: Sinauer Associates, 1993).

33 The Milkman quotation initiates his chapter 6 in *Perspectives on Evolution* (Sunderland, Mass.: Sinauer Associates, 1982). There are many works that discuss the possibility of natural selection working at higher levels of organization than the individual. For an introduction, I recommend chapter 3 of my *Natural Selection* (New York: Oxford University Press, 1992).

34–35 The history of genetics is nicely presented in Elof A. Carlson, *The Gene: A Critical History* (Iowa City: Iowa State University Press, 1989). Watson and Crick's classic publication is "Molecular structure of nucleic acids," in *Nature* 171 (1953):737–38.

37–38 A good introduction to population genetics is J. Maynard Smith, *Evolutionary Genetics* (Oxford: Oxford University Press, 1989). Russell Lande's "Natural selection and random genetic drift in phenotypic evolution," *Evolution* 30 (1976): 314–34 remains the best detailed consideration of possible rates of change in quantitative characters, such as horses' coat colors. The possibility of rapid evolution of a complete vertebrate eye is discussed on p. 78 of Richard Dawkins's *River Out of Eden* (New York: Basic Books, 1995).

3: DESIGN FOR WHAT?
•••••••

39 The quotation from Aristotle is from p. 103 of *Aristotle: Parts of Animals*, trans. A. L. Peck (Cambridge, Mass.: Harvard University Press, 1955). The textbook was Tracy I. Storer's *General Zoology* (New York and London: McGraw-Hill, 1943).

43 The concepts of inclusive fitness and kin selection were first elucidated by William D. Hamilton in "The genetical evolution of social behavior," *Journal of Theoretical Biology*, 7 (1964): 1–52.

45–47 The most comprehensive and authoritative work on social insects remains E. O. Wilson's *The Insect Societies* (Cambridge, Mass.: Harvard University Press, 1971). An excellent recent work is that of Andrew F. G.

Bourke and N. R. Franks, *Social Evolution in Ants* (Princeton, N.J.: Princeton University Press, 1995). Darwin discussed social insects, and expressed the quoted qualm, in chapter 8 of *The Origin of Species*. See also Helena Cronin's comments on pp. 298–99 of *The Ant and the Peacock* (Cambridge: Cambridge University Press, 1991).

50 The superorganism concept is nicely presented by Thomas D. Seeley, "The honey bee colony as a superorganism," *American Scientist* 77 (1989): 546–53. I can also recommend the lavishly illustrated and less technical coverage of the same material in *The Honey Bee,* by James L. Gould and Carol G. Gould (New York: Scientific American Library, 1988).

53–55 The classic work on the use of game theory for evolutionary problems is J. Maynard Smith's *Evolution and the Theory of Games* (New York: Cambridge University Press, 1982). The classic work on sex ratio is Eric L. Charnov's *The Theory of Sex Allocation* (Princeton, N.J.: Princeton University Press, 1982). Seal reproduction is discussed on pp. 106–10 of Richard Dawkins's *River Out of Eden* (New York: Basic Books, 1995).

57–58 The tendency for animal groupings to be of greater than optimal size is pointed out by R. M. Sibly, "Optimal group size is unstable," *Animal Behavior* 31 (1983): 947–48 and by M. Higashi and N. Yamamura, "What determines animal group size? Insider-outsider conflict and its resolution," *American Naturalist* 142 (1993): 553–63.

4: THE ADAPTIVE BODY
••••••

59 Dawkins's discussion of blueprints and recipes is on p. 175 of *The Extended Phenotype* (Oxford and San Francisco: Freeman, 1982) and on pp. 294–98 of *The Blind Watchmaker* (New York and London: Norton, 1986).

60 A good introduction to the molecular mechanisms and other general problems of development, with emphasis on the way organisms make use of spontaneously occurring processes, is Stuart Kauffman, *The Origins of Order* (New York: Oxford University Press, 1983).

62 The Maynard Smith quotation is from chapter 9 of his *The Problems of Biology* (Oxford: Oxford University Press, 1986).

63–65 General biology textbooks in mid-century always discussed the vitalism-mechanism controversy, but today's authors seem never to have heard of vitalism. Detailed treatments are available from works by historians or philosophers, for instance, pp. 22–24 of Elliott Sober's *Philosophy of Biology* (Boulder, Colo.: Westview Press, 1993). The animal-mind concept and its relevance to biology are effectively championed in several works by D. R. Griffin, most recently in his *Animal Minds* (Chicago: University of Chicago Press, 1992). My views on mentalism as a form of vitalism are developed in more detail on pp. 21–24 of *Oxford Surveys in Evolutionary Biology*, vol. 2 (1985), and on pp. 3–5 of my *Natural Selection* (New York: Oxford University Press, 1992). I doubt the reality of simian capabilities for symbolic language, for reasons given by Steven Pinker on pp. 337–42 of his *The Language Instinct* (New York: Morrow, 1994).

66 The textbook by Carlson and Johnson is *The Machinery of the Body* (New York: Morrow, several editions, 1938–53).

67 Mechanisms of muscle contraction and nerve conduction are given in greater detail in many excellent general textbooks, for instance, pp. 824–34 of William K. Purves, Gordon H. Orians, and H. Craig Heller, *Life: The Science of Biology* (Sunderland, Mass.: Sinauer Associates, 1992). R. McNeill Alexander's *The Human Machine* (New York: Columbia University Press, 1992) is an informative and readable account of the mechanical actions of the human body. See esp. chapter 2, "Handling."

68–70 The dearth of shared descriptors is the essence of the mind-brain problem widely discussed by philosophers. A clear introduction to this topic is provided by P. M. Churchland in *Matter and Consciousness* (Cambridge, Mass.: MIT Press, 1988). Data on Mr. Bloodthirsty are from *The Guinness Book of Records* (New York: Facts on File, 1994), p. 63. The George Liles quotation is from "Why is life so complex?," *MBL Science*, 3:9–13.

71–72 The metabolic roles of mitochondria are standard text-book materials. Pp. 74–75 and 152–55 of W. K. Purves and collaborators' *Life* (Sunderland, Mass.: Sinauer Associates, 1995) give a clear and current account.

73–74 The many forms of conflict and cooperation within a cell are reviewed by Laurence D. Hurst and collabora-tors, "Genetic conflicts," *The Quarterly Review of Biol-ogy* 71 (1996, in press).

5: WHAT USE IS SEX?
······

76 Maynard Smith's article is in *Journal of Theoretical Biology* 30 (1971): 319–35. The book is his *The Evolu-tion of Sex* (London: Cambridge University Press, 1978).

77–79 The origin of sexuality in relation to genetic proofread-ing and other factors is discussed in *The Evolution of Sex,* ed. R. E. Michod and B. R. Levin (Sunderland, Mass: Sinauer Associates, 1988). See esp. chaps. 2, 3, 9, and 12.

81–82 The classic article on the origin of the sperm-egg distinc-tion is by G. A. Parker and collaborators, "The origin and evolution of gamete dimorphism and the male-female phenomenon," *Journal of Theoretical Biology* 36 (1972): 529–53.

85 Detailed discussions of the advantages of parthenogene-sis are provided in Michod and Levin, *The Evolution of Sex,* chapters 1 and 17. *Why Be an Hermaphrodite?* by E. L. Charnov, J. Maynard Smith, and J. J. Bull is in *Nature* 263 (1976): 125–26.

87 The size-advantage model was originally elaborated by Michael T. Ghiselin, "The evolution of hermaphroditism among animals," *The Quarterly Review of Biology* 44 (1969): 189–208. It is discussed further in many modern works on evolution.

89–90 The best detailed work on sex ratio theory and related topics is that of E. L. Charnov, *The Theory of Sex Allo-cation* (Princeton, N.J.: Princeton University Press, 1982).

91–93 The best review of sexual size dimorphism and the dwarf-male phenomenon is that of Michael T. Ghiselin in *The Economy of Nature and the Evolution of Sex*

(Berkeley: University of California Press, 1974), esp. pp. 193–212.

6: THE HUMAN EXPERIENCE OF SEX AND REPRODUCTION
• • • • • • •

95–101 The classic work on weaning conflict, and on conflict between relatives generally, is R. L. Trivers's "Parent-Offspring Conflict," *American Zoologist* 14 (1974): 249–64. David Haig's work is "Genetic Conflicts in Human Pregnancy," *The Quarterly Review of Biology* 68 (1993): 495–532.

102 Margie Profet's idea was first outlined in "The evolution of pregnancy sickness as protection to the embryo against Pleistocene teratogens," *Evolutionary Theory* 8 (1988): 177–90. A fuller presentation is chapter 8 of *The Adapted Mind*, ed. J. H. Barkow, L. Cosmides, and J. Tooby (New York: Oxford University Press, 1992). Her *Protecting Your Baby-to-Be: Preventing Birth Defects in the First Three Months of Pregnancy* was published in 1995 (Reading, Mass.: Addison-Wesley).

111 Symons's book is published by Oxford University Press, Hrdy's by Harvard University Press. Each author wrote a constructively critical review of the other's work in *The Quarterly Review of Biology* 54 (1980): 309–14 and 57 (1982): 297–300. Informative recent discussions and introductions to prior literature on factors that influence mate selection in diverse human societies are provided by Randy Thornhill and collaborators in "Human female orgasm and mate fluctuating asymmetry," *Animal Behavior* 50 (1995):1601–15 and by David Buss in chapter 5 of Barkow, Cosmides, and Tooby, *The Adapted Mind*. See also the update of Symons's ideas on pp. 91–124 of *Theories of Human Sexuality*, ed. J. H. Geer and W. T. O'Donohue (New York: Plenum, 1987).

7: OLD AGE AND OTHER CURSES
• • • • • • •

112–17 Two well-reasoned and comprehensive treatments of senescence are M. R. Rose's *Evolutionary Biology of*

Aging (New York: Oxford University Press, 1991) and Caleb Finch's *Longevity, Senescence, and the Genome* (Chicago: University of Chicago Press, 1991).

118   An admirably clear statement of the effectiveness of selection in relation to age was presented in 1952 by the British Nobel laureate P. B. Medawar in *An Unsolved Problem in Biology* (London: H. K. Lewis). A rigorous mathematical development of the theory was provided by William D. Hamilton, "The moulding of senescence by natural selection," *The Journal of Theoretical Biology* 12 (1966): 12–45. Hamilton was the first to make explicit the effectiveness of selection as proportional to the product of survivorship and reproductive value. He also showed that eternal youth in any population would inevitably be unstable. Only an exponentially increasing fertility would avoid the evolution of senescence. For a fuller discussion of difficulties in the measurement of senescence, see G. C. Williams and P. D. Taylor's chapter (pp. 235–45) in *The Evolution of Longevity in Animals,* ed. A. D. Woodhead and K. H. Thompson (New York and London: Plenum, 1987).

R. A. Fisher introduced the concept of reproductive value in chapter 2 of his classic *The Genetical Theory of Natural Selection* (Oxford: Clarendon Press, 1930). In the interest of conceptual simplicity, I used some obviously inexact approximations in this chapter, such as instantaneous attainment of full fertility at puberty. Fisher calculated actual values from fertility and survival rates for Australian women of the year 1911. His curve, like mine, started with a value of about 2 at age zero. Unlike my hypothetical Stone Age population, the Australian was growing rapidly and had far lower mortality at every age. So his curve peaked at about 2.8 at age eighteen and then declined to 0 at menopause at about fifty. For use in the understanding of senescence, Fisher's *reproductive value*, an estimated number of babies to be produced, should be replaced by what might be called *genetic value*, an estimate of capability for getting one's own genes into future generations. Postmenopausal women may have appreciable genetic value if they gather food to be shared with relatives, or help them in any other way.

119 Rates of survival of infants to maturity in the Stone Age, or even in modern tribal societies, can obviously not be estimated with any precision. An important recent work on this topic is *Hunter-Gatherer Demography, Past and Present*, ed. Betty Meehan and Neville White (Sydney, Australia: University of Sydney Press, 1990). Neonatal mortality is usually underestimated, especially that from infanticide, recently recognized as important. See *Infanticide: Comparative and Evolutionary Perspectives*, ed. Glenn Hausfater and Sarah B. Hrdy (New York: Aldine, 1984). For more recent discussion, see Hrdy and collaborators in "Infanticide: Let's not throw out the baby with the bath water," *Evolutionary Anthropology* 3 (1995): 149–51.

124 Cannon's book was published in New York by Norton, Estabrook's in New York by Macmillan.

131 I was led to the quotation from *The African Queen* by Dr. John Hartung, State University Health Sciences Center, Brooklyn, N.Y.

8: MEDICAL IMPLICATIONS
•••••••

Medical implications of modern Darwinism are discussed in more (but grossly inadequate) detail by Randolph M. Nesse and George C. Williams in *Evolution and Healing: The New Science of Darwinian Medicine* (London: Weidenfeld & Nicolson, 1995). This book is published in the United States as *Why We Get Sick* (New York: Times Books, 1995). Several other recent, general-audience works also take a Darwinian view of medical problems: H. Boyd Eaton and collaborators, *The Paleolithic Prescription* (New York: Harper & Row, 1988); Paul W. Ewald, *Evolution of Infectious Disease* (New York: Oxford University Press, 1994); Margie Profet, *Protecting Your Baby-to-Be: Preventing Birth Defects in the First Three Months of Pregnancy* (Reading, Mass.: Addison-Wesley, 1995); Marc Lappé, *Evolutionary Medicine: Rethinking the Origins of Disease* (San Francisco: Sierra Club Books, 1994).

134–40 Early vertebrate evolution is discussed in many textbooks. A good introduction is chapter 2 of Arnold G.

Kluge, *Chordate Structure and Function*, 2nd ed. (New York: Macmillan, 1977).

142   Ewald's work was published in *Journal of Theoretical Biology* 86 (1980): 169–76.

9: PHILOSOPHICAL IMPLICATIONS
••••••

152   The Tennyson quotation is from Canto 55 of "In Memoriam." That from Lewis Carroll is from chapter 5 of *Through the Looking Glass.*

153–54   The Shaw quotation is from the section "The Moment and the Man" in the preface to his *Back to Methuselah.* The Huxley quotation is from p. 84 of his 1893 address, as reprinted by Princeton University Press in *Evolution and Ethics*, James Paradis and G. C. Williams, editors (1989). The Dawkins quotation is from p. 215 of his *The Selfish Gene* (New York: Oxford University Press, 1976). I provide more detail on the wickedness of biological phenomena and prospects for morality in an immoral world on pp. 179–214 of Paradis and Williams, *Evolution and Ethics.*

155–56   For data on infanticide in human and animal populations, see Sarah B. Hrdy, *The Langurs of Abu* (Cambridge, Mass.: Harvard University Press, 1977); Glenn Hausfater and Sarah B. Hrdy, *Infanticide: Comparative and Evolutionary Perspectives* (New York: Aldine, 1984).

157   The Lavick quotation is from p. 3 of Hrdy, *The Langurs of Abu.* Her article "Infanticide as a primate reproductive strategy" is in *American Scientist* 65: 40–49.

158–60   The complexities of conception are discussed by R. J. Aitkin, "The complications of conception," *Science* 269 (1995): 39 and by other articles in the same issue. The problem of fetal rights is discussed realistically by Bentley Glass in *The Quarterly Review of Biology* 67 (1992): 501–4; H. J. Morowitz and J. S. Trefil, *The Facts of Life* (New York: Oxford University Press, 1992); and James Rachels, *Created from Animals* (Oxford: Oxford University Press, 1990). Garrett Hardin's "The Meaninglessness of the Word Protoplasm," *Scientific Monthly* 82 (1956): 112–20, was a decisive event in the history of this ill-starred term. For an introduction to the literature on individuals of mixed genotype, see the article by Shigeki Uehara and collaborators in *Fertility & Sterility* 63 (1995): 189–92.

160 E. O. Wilson, *Sociobiology: The New Synthesis* (Cambridge, Mass.: Belknap Press of Harvard University Press, 1975).

161 D. S. Wilson and collaborators, "Species of thought: A commentary on evolutionary epistemology," *Biology and Philosophy* 5 (1990): 37–62.

162 For a review of Cosmides and Tooby's and related work, see J. H. Barkow, Leda Cosmides, and John Tooby, eds., *The Adapted Mind* (New York: Oxford University Press, 1992).

163 Paul Davies, *About Time* (New York: Simon & Schuster, 1995).

164 The distinguished historian is R. J. Richards, and the "lubricious slide" quotation is from p. 73 of his *Darwin and the Emergence of Evolutionary Theories of Mind and Behavior* (Chicago and London: University of Chicago Press, 1987). David Hume's complaint about deriving *ought* from *is* is quoted at length on p. 87 of Michael Ruse's *Taking Darwin Seriously* (New York and Oxford: Basil Blackwell, 1986). Current thinking on the mental-material domain mix is explained with admiral clarity in Paul Churchland's *Matter and Consciousness* (Cambridge, Mass.: MIT Press, 1988). For more on the uses of codical and material domains in biology, see my *Natural Selection* (Princeton, N.J.: Princeton University Press, 1992), esp. pp. 10–16.